Equipment and Components in the Oil and Gas Industry Volume 2

Equipment and Components in the Oil and Gas Industry Volume 2: Components provides an overview of the components used in the oil and gas industry, including instrumentation, pipe components, and safety components. Using practical industry examples and an accessible approach, the book is a key reference point for those seeking to learn more about the industry.

Covering both larger and smaller components used throughout the oil and gas industry, the book details the theory behind pressure gauges, temperature gauges, flow gauges, and level gauges. It then goes on to discuss piping components, such as pipes, flanges, and gaskets and introduces piping special components. Valves are particularly crucial to the oil and gas industry, including on/off valves, control valves, safety valves, and special valves. The book also details actuators, sprinklers, fire and gas detectors, hoses, and hose reels, along with electrical components such as switches, cables, wires, and cable glands. Finally, the book ends with a discussion of heating, ventilation, and air conditioning (HVAC) components.

This book will be of interest to mechanical and chemical engineers working in the oil and gas industry.

Equipment and Components in the Oil and Gas Industry Volume 2

Components

Karan Sotoodeh

CRC Press is an imprint of the
Taylor & Francis Group, an **informa** business

Designed cover image: Shutterstock

First edition published 2024
by CRC Press
2385 NW Executive Center Drive, Suite 320, Boca Raton FL 33431

and by CRC Press
4 Park Square, Milton Park, Abingdon, Oxon, OX14 4RN

CRC Press is an imprint of Taylor & Francis Group, LLC

Library of Congress Cataloging-in-Publication Data
Names: Sotoodeh, Karan, author.
Title: Equipment and components in the oil and gas industry / Karan Sotoodeh.
Description: Boca Raton : CRC Press, 2024. | Includes bibliographical references and index.
Identifiers: LCCN 2023051960 (print) | LCCN 2023051961 (ebook) | ISBN 9781032739076 (hbk ; volume 1) | ISBN 9781032739991 (pbk ; volume 1) | ISBN 9781032731476 (hbk ; volume 2) | ISBN 9781032737799 (pbk ; volume 2) | ISBN 9781003465881 (ebk ; volume 2) | ISBN 9781003467151 (ebk ; volume 1)
Subjects: LCSH: Petroleum engineering—Equipment and supplies.
Classification: LCC TN871.5 .S626 2024 (print) | LCC TN871.5 (ebook) | DDC 681/.7665—dc23/eng/20231122
LC record available at https://lccn.loc.gov/2023051960
LC ebook record available at https://lccn.loc.gov/2023051961

ISBN: 978-1-032-73147-6 (hbk)
ISBN: 978-1-032-73779-9 (pbk)
ISBN: 978-1-003-46588-1 (ebk)

DOI: 10.1201/9781003465881

Typeset in Times
by Apex CoVantage, LLC

Contents

Preface

It is the purpose of this book to provide a general overview of oil and gas equipment and components. Volume 2 focuses on the main and most important components. Instruments, piping, valves and actuators, safety, flare, and electrical components are among the main components discussed in this volume of the book. Chapter 1 discusses the instruments used in the oil, gas, and petrochemical industries to monitor and control their processes. A plant's instrumentation ensures that the plant operates within defined parameters in order to produce materials of consistent quality. Generally, instrumentation can be defined as the art and science of measuring and controlling. A brief introduction to measurement and control terminology is provided in this chapter. There is a discussion in Chapter 2 of piping, pipe fittings, pipe connections, and piping special items. Pipelines and piping are both used to transport liquids, gases, and slurries. A piping system is usually connected to a variety of equipment and carries fluids that are to be processed within that equipment within the framework of a complex network. In a piping system, a valve is responsible for stopping or starting fluid flow. In addition, it is used to regulate and control fluid flow, to prevent backflow, and to ensure safety. As part of a piping system, it plays a crucial role. Valve costs account for approximately 20% to 30% of the total cost of piping. It is equally important for the success of a process plant to understand how valves operate, are maintained, and are adjusted. As a result, Chapter 3 is entirely devoted to industrial valves. A valve must be able to be operated. All valves can be operated manually or by an actuator, except pressure relief valves and check valves. To increase efficiency and productivity in the oil and gas industry, valves with actuators have become increasingly necessary. As discussed in Chapter 4, actuators are devices that convert external energy sources into mechanical motions, such as air, hydraulic power, or electricity. Components and systems essential to safety are discussed in Chapter 5. The purpose of this chapter is to provide a detailed overview of the essential SCEs as well as the associated equipment and facilities, including emergency shutdown systems (ESDs), fire and gas detection systems (FGDS), flare monitoring systems, safety relief and blowdown systems, and firefighting systems. In addition to sprinklers, fire and gas detectors, fire hydrants, and so on, this chapter also provides detailed explanations of the basic principles of fire prevention and control. The flare system discussed in Chapter 6 plays a vital role in the safe and efficient operation of a process plant because it is an integral part of the emergency relief of flammable substances in liquid or gaseous phases. Additionally, this chapter provides information regarding flare design considerations, flare system operating principles, flare terminology, and noise and smoke problems and solutions. Last, Chapter 7 discusses the essential components of electrical systems.

Author

Karan Sotoodeh, an Iranian author and engineer, formerly worked for Baker Hughes in his last position as a senior–lead valve and actuator engineer in the subsea oil and gas industry. He earned a PhD in safety and reliability in mechanical engineering at the University of Stavanger in 2021. Karan Sotoodeh has almost two decades of experience in the oil and gas industry, mainly with valves, piping, actuators, and material engineering. He has written 11 books about piping, valves, actuators, corrosion, and material selection and approximately 50 papers in peer-reviewed journals. Dr. Sotoodeh has also been selected for international conferences in the United States, Germany, and China to talk about valves, actuators, and piping. He has worked with many valve suppliers in the United Kingdom, Italy, France, Germany, and Norway. He loves traveling, running, swimming, and spending time in nature.

1 Instrumentation

1.1 INTRODUCTION

The oil, gas, and petrochemical industries use instrumentation to monitor and control their process plants. A plant's instrumentation ensures that the plant operates within defined parameters in order to produce materials of consistent quality and within the required specifications. Furthermore, it ensures that the plant is operated in a safe manner and operates to correct out-of-tolerance operation and to shut down the plant in a timely manner in order to avoid hazardous conditions. Instrumentation consists of sensor elements, signal transmitters, controllers, indicators and alarms, actuated valves, logic circuits, and user interfaces. A process instrument is a device that measures process parameters such as pressure, temperature, liquid level, flow, velocity, composition, density, and weight. It is necessary to install reliable flow, pressure, and temperature measurement and gas analysis equipment in order to conduct oil and gas drilling, extraction, and production. The reliability of processes and the longevity of offshore and onshore platforms are enhanced through optimal control of equipment, even in extreme environments. By choosing the right instrumentation, you can help protect the environment by detecting leaks or stopping them, as well as making transport and storage of oil and gas more reliable. To produce and deliver a product to the market, modern processing plants use a number of control loop networks. Control loops, including instrument sensors and transmitters, ensure that process variables, such as pressure, temperature, level, flow, and so on, are maintained within the required operating ranges. According to Figure 1.1, each of these loops receives variations from the desired quality and changes the process variable accordingly. In order to reduce the effect of process variations on the desired set point, sensors collect information from the process variable, and transmitters transmit that information to the controller (e.g. the control room). In order to return the process variable to the desired state, a controller processes the information and determines the actions that need to be taken. Control valves are the most common final control element in process control. The control valve is a type of instrument valve that regulates fluid flow by altering the passage size and regulating the process variable as closely as possible to the desired value. In the third chapter of this book, control valves are discussed in more detail.

1.2 INSTRUMENTATION AND MEASUREMENT SYSTEMS

Instrumentation is defined as "the art and science of measuring and controlling". The term "instrumentation" can refer to a field in which instrument technicians and engineers work, or it can refer to the methods and purposes for which

DOI: 10.1201/9781003465881-1

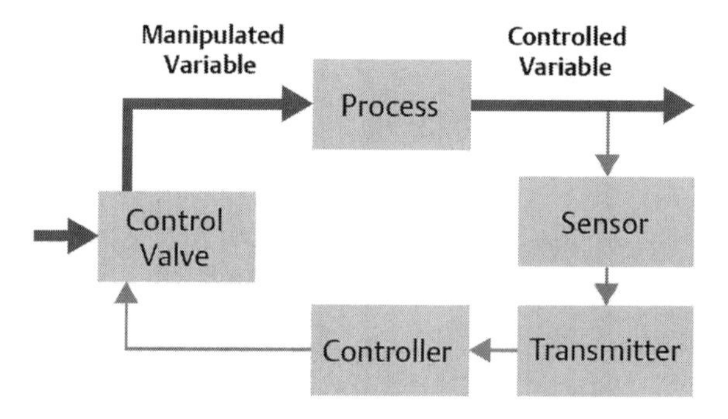

FIGURE 1.1 A control loop.

instruments are used. Instruments are devices that transform physical variables of interest (the measurand) into a form suitable for recording (the measurement).

Physicist Lord Kelvin eloquently expressed the importance of measurement with the following statement: "I often say that when you can measure what you are speaking about and can express it in numbers, you know something about it; when you cannot express it in numbers your knowledge is of a meager and unsatisfactory kind." Since the dawn of human civilization, measurement techniques have played a crucial role in regulating the transfer of goods in barter trade to ensure that exchanges are fair. As a result of the Industrial Revolution in the 19th century, new instruments and measurement techniques were developed rapidly in order to meet the demands of industrialized production. New industrial technologies have grown rapidly since that time. As a result of numerous developments in electronics in general and in computers in particular in the latter part of the 20th century, this has been particularly evident. Consequently, a parallel growth in instruments and measurement techniques has been necessary. Measurement involves the use of instruments to determine quantities or variables. It is common to refer to the measuring instrument as a measurement system due to its modular nature. The structure of measurement system is illustrated in Figure 1.2.

Following is a summary of the basic measurement system:

1. Transducers consist of a sensing element and a signal conditioning element (variable conversion element). Sensors detect physical parameters, and the variable conversion element converts them into electrical signals. The variable conversion element is not always necessary when sensors can perform the sensing and converting.
2. A signal processing element converts small electrical quantities into quantities that can be used by the rest of the system.
3. When the final display step is performed, the signal presentation is used to display the measured quantity's reading.

A measuring system begins with the primary *sensor*. A sensor is a device that detects changes in the environment and responds to outputs from another system.

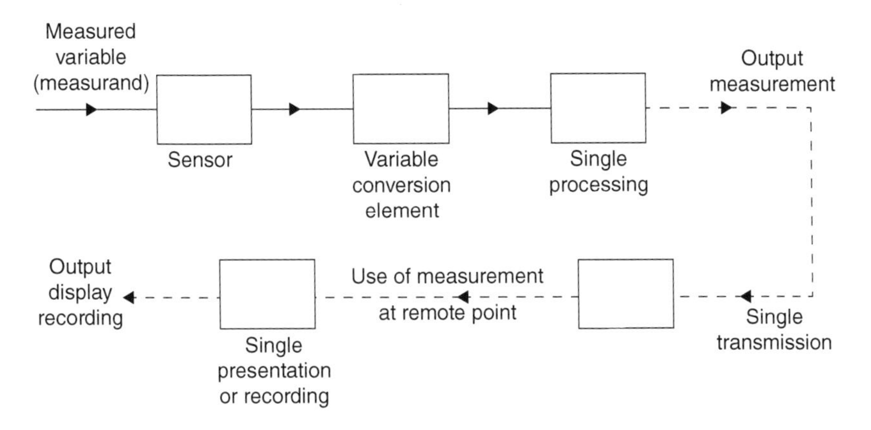

FIGURE 1.2 Structure of measurement system.

The purpose of a sensor is to convert a physical phenomenon into a measurable analog voltage (or sometimes a digital signal), which is then displayed on a human-readable display or transmitted to be analyzed. The microphone is one of the most well-known sensors. It converts sound energy into an electrical signal that can be amplified, transmitted, recorded, and reproduced. The use of sensors is ubiquitous in our daily lives. Traditionally, mercury thermometers have been used for measuring temperature for a very long time. Based on the fact that colored mercury reacts consistently and linearly to temperature changes in a closed tube, this method uses colored mercury in a closed tube.

It is necessary to include *variable conversion elements* when the output variable of a primary transducer is in an inconvenient form and has to be converted into a more convenient format. As an example, a displacement measuring strain gauge produces a varying resistance as its output. It is possible to measure applied force, pressure, torque, and so on using strain gauges by converting them into electrical signals. A strain gauge measures a change in electrical resistance as a result of force causing strain. Due to the difficulty of measuring resistance changes, they are converted into voltage changes using a bridge circuit, which is a typical variable conversion element. Occasionally, the primary sensor and variable conversion element are combined; this is known as a *transducer.*

It is possible to improve the quality of the output of a measurement system by means of signal processing elements. An electronic amplifier is a type of signal processing element that amplifies the output of the primary transducer or variable conversion element, thereby improving the sensitivity and resolution of the measurement. When the primary transducer has a low output, this element of a measurement system is particularly important. The output of thermocouples, for instance, is typically only a few millivolts. There are also signal processing elements that filter out induced noise and remove mean levels. Signal processing is sometimes incorporated into a transducer, which is then referred to as a *transmitter.*

A measurement system may also contain one or two other components, one for *transmitting the signal* to a remote location and the other for displaying or recording the signal if it is not fed automatically into a feedback control system. Transmission of signals is required when the observation or application point of the output of a measurement system is some distance away from the site of the primary transducer. Occasionally, this separation is made solely for convenience, but most often it results from inaccessibility or environmental conditions unsuitable for mounting the signal presentation/recording device at the site of the primary transducer. Traditionally, signal transmission has been accomplished by single or multi-cored cables, which are often screened in order to avoid signal corruption caused by electrical noise. Modern installations are increasingly using fiber-optic cables, in part due to their low transmission loss and resistance to the effects of magnetic and electrical fields.

A measurement system's final optional element is the point at which the measured signal is used. There are instances when this element is omitted entirely since the measurement is part of an automatic control scheme, and the transmitted signal is directly fed into the control system. Alternatively, the signal presentation unit or the signal recording unit may be used in this element of a measurement system. Based on the requirements of the particular measurement application, these can take a variety of forms.

1.3 MEASUREMENT AND CONTROL TERMINOLOGY

Feedback control: Feedback control involves measuring the variable being controlled and comparing it with a target value. The difference between the actual value and the desired value is known as the error. In order to minimize this error, feedback control manipulates an input to the system. A feedback control system reacts to the system and attempts to minimize this error. It is generally possible to enter the desired output into the system through a user interface. In this process, the output of the system is measured (using a flow meter, thermometer, or similar instrument), and a difference is calculated. As a result of this difference, the system inputs are controlled to reduce the error in the system. In order to understand the principle of feedback control, consider an electric oven. In order to bake cookies, one must preheat the oven to 350°F. A sensor inside the oven takes a reading after the desired temperature has been set. If the oven temperature is below the set temperature, a signal is sent to the heater to activate until the oven reaches the set temperature. A variable to be controlled (oven temperature) is measured to determine how the input variable (heat into oven) should be manipulated in order to achieve the desired result. Human behavior can also be used to demonstrate feedback control. For example, a person who goes outside in winter will experience a drop in body temperature. In response to this signal, the brain (controller) produces a motor action in order to put on a jacket. The result is a reduction in the discrepancy between the skin temperature and the physiological set point of the individual.

In Figure 1.3, a very applicable example of a control system is the level control system for a pressure vessel. The level transmitter (LT) detects the level of liquid or gas inside the pressure vessel and sends the signal to the level controller (LC), which in this case acts as a feedback control. Whenever the liquid level in the tank falls below the target value, the level controller commands the control valve to adjust (closing more) so that the liquid level in the tank is raised. In contrast, if the liquid level in the pressure vessel is above the target level, the level controller (feedback control) opens the valve, allowing the liquid level in the tank to decrease. For controlling the flow into the pressure vessel, a flow transmitter, controller, and control valve are also installed before the vessel. A flow controller is also considered a feedback control.

Single loop control: The combination of a flow transmitter, a flow controller, and a control valve is referred to as a single loop control. Therefore, there are two single loop controls shown in Figure 1.3, one for flow control and one for level control.

Closed-loop system: Closed-loop control systems adjust the input variable to minimize the error between the measured output variable and its set point. This is synonymous with feedback control, which involves feeding back deviations between the measured variable and a set point in order to generate the desired control action.

Open-loop system: Open-loop control refers to a control system that does not use feedback information to adjust the process. In an open-loop controller, one or more measured variables are used to generate control actions based on existing equations or models. Open-loop controls, also referred to as non-feedback controls, are continuous control systems in which the output has no effect on the control action of the input signal. Therefore, in an open-loop control system, the output is neither measured nor fed back for comparison with the input. A traditional toaster is an example of an open-loop control system. The user of a toaster

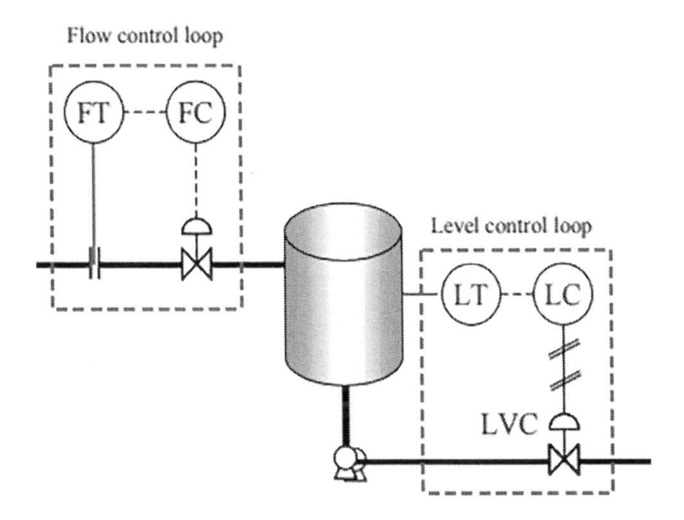

FIGURE 1.3 Flow and level feedback control system.

is only aware of when it begins and finishes toasting, with no control over the input (two slices of bread) nor the process (time, temperature, etc.). Various open-loop controls are available, such as on/off switches for valves, machinery, lights, motors and heaters, where the control result is known to be approximately sufficient under normal conditions without requiring feedback. It is advantageous to use open-loop control in these cases because it reduces the number and complexity of components. In contrast to closed-loop systems, open-loop systems cannot correct errors they make or compensate for outside disturbances, nor can they engage in machine learning.

Cascade control: The cascade control arrangement involves two (or more) controllers, each of whose output drives the set point of the other controller. In this case, the level controller drives the set point of the flow controller in order to maintain the level at its set point (see Figure 1.4). As a result, the flow controller drives a control valve to match the flow with the set point requested by the level controller.

Feedforward control: The goal of feedforward control is to reject disturbances that are persistent and cannot be adequately rejected by feedback control. A feedforward control system is generally used in conjunction with a feedback control system and is not usually implemented on its own. The application of feedforward control can enhance the performance of control systems in certain circumstances. In feedforward control, the basic concept is to measure important disturbance variables and correct them before they result in a process disruption. There are two key characteristics that must be present:

1. Despite the attempts of a feedback control system to regulate the effects of an identifiable disturbance on the measured variable, this disturbance significantly affects the measurement
2. It is possible to measure this disturbance, perhaps with the addition of instrumentation

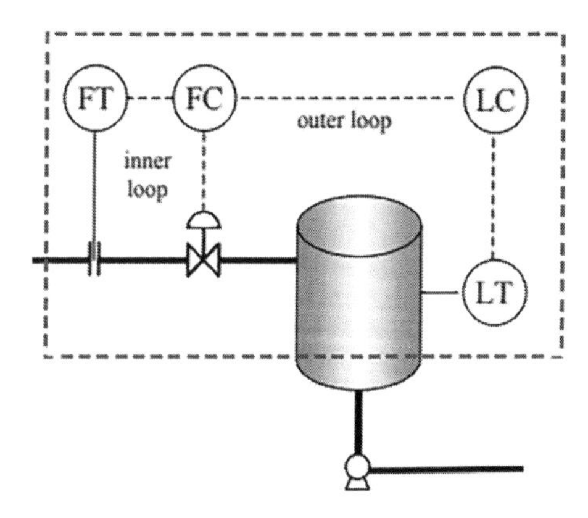

FIGURE 1.4 Flow and level cascade control system.

Note: A control system can be configured in three different ways: feedback control, cascade control, and feedforward control.

Controller: A device used to compute the amount of change that needs to be made at the final control element.

Final control element: A device used to perform the change instructed by the controller.

Process: A mass that changes its characteristics due to chemical reaction or quantity, such as volume expansion. Examples of processes are liquid flow, liquid level, and so on.

Transmitter: A device used to amplify the signal from a sensor and relay it to the controller.

Sensor: A device used to measure the changes in process. A process variable whose value should be maintained. As an example, if a process temperature needs to be maintained within 5°C of 100°C, then the set point temperature is 100°C.

Manipulated variable: The variable that is being changed by the final control element. An example of a control valve (final control element) is used to adjust the amount of fluid entering the tank. In this case, the manipulated variable is the amount of water entering the tank. Despite the change in output demand, the water flow is manipulated to keep the level constant.

Controlled variable: The variable that is being controlled. During an experiment to examine the growth of a plant, the temperature could be considered a control variable if it is controlled. There are a number of other variables that could serve as control variables, including the amount of light, the duration of the experiment, the amount of water, and the pot of the plant.

Load variable: Process load refers to the set of all parameters, excluding the controlled variable.

Measured variable: The variable that is being measured by the sensor.

1.4 INSTRUMENTATION CLASSIFICATIONS

The following criteria can be used to subdivide instruments into different classes:

Passive or ***active***: In terms of power consumption, we can divide instruments into active and passive categories. An active instrument requires power to function, either from a battery or an external voltage source. "Passive" simply refers to an instrument in which no power is necessary to operate its components. Passive instruments do not contain any electrical power source. In self-generating (or passive) instruments, the energy requirements are entirely met by the input signal. Alternatively, power-operated (or active) instruments require additional power to operate, such as compressed air, electricity, hydraulic supply, and so on.

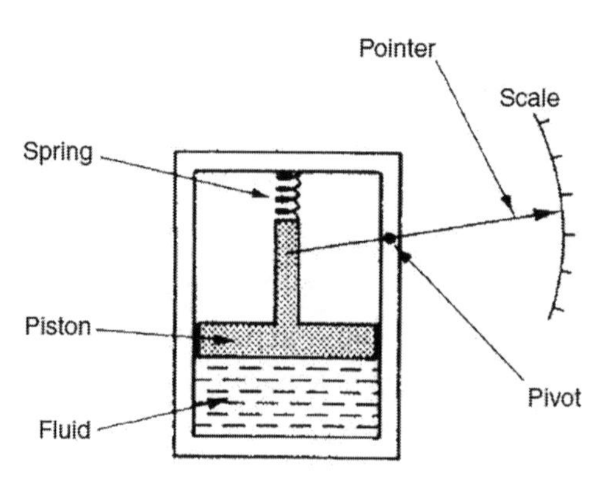

FIGURE 1.5 A passive pressure gauge.

A power source is contained in the active instrument. It is usually electric power that powers active instruments, but there are circumstances in which other forms of energy, such as pneumatics or hydraulics, may be used. Passive instruments are cheaper, but they have better resolution. Active instruments are powered by an external power source. An example of a passive instrument is a pressure-measuring device, illustrated in Figure 1.5. By translating the pressure of the fluid into a movement of the pointer against the scale, the pressure of the fluid can be measured. There is no other source of energy that is added to the system in order to move the pointer: the energy is derived exclusively from the change in pressure measured. A petrol tank level indicator (see Figure 1.6) is an example of active instrument. It consists of a potentiometer arm that moves with a change in petrol level and an output signal that is a proportion of the voltage source applied across the two ends of the potentiometer. By modulating the voltage from this external power source, the primary transducer float system is simply changing the value of the voltage. An external power source provides the energy in the output signal.

Contacting or ***noncontacting type***: An instrument that is in contact with the medium is considered a contacting instrument. A clinical thermometer is an example of such an instrument. Conversely, there are instruments that are noncontact or proximity in nature. It is possible for these instruments to measure the desired input even if they are not in close contact with the measuring medium. Optical pyrometers, for example, monitor the temperature of a blast furnace without coming into contact with it. With a contact-type instrument, it is not possible to measure the temperature of a highly heated body. The optical pyrometer measures temperature by measuring how much light of a certain wavelength is emitted by a hot object. The pyrometer works by matching the brightness of an object to the brightness of the filament contained inside. Temperatures of furnaces, molten metals, and other overheated materials or liquids can be measured with the optical pyrometer.

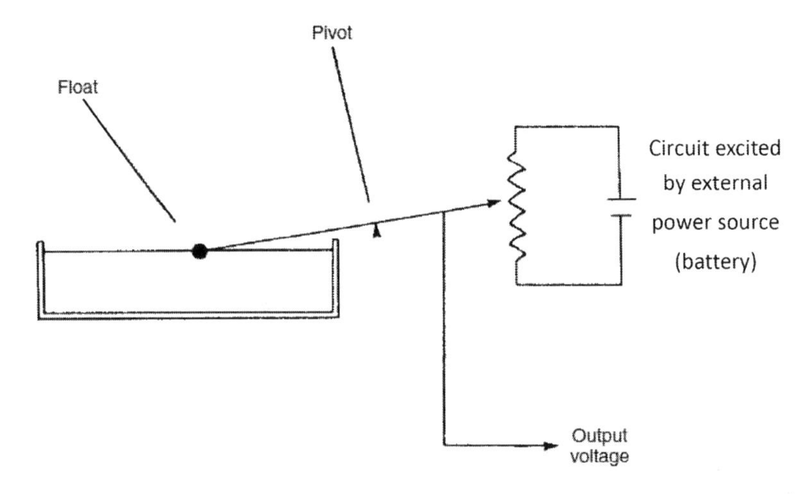

FIGURE 1.6 Float-type petrol tank level indicator (active measurement).

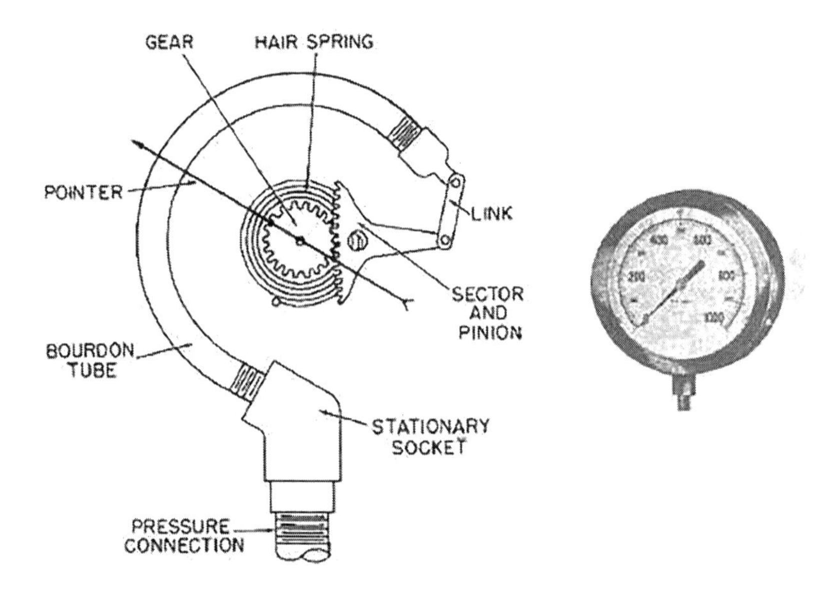

FIGURE 1.7 An analog pressure gauge.

Analog or ***digital type:*** A digital instrument displays the output as text or numbers on a digital display screen, in contrast to an analog instrument, which displays the output as a deflection of a pointer on a scale (see an analog pressure gauge in Figure 1.7). Generally, analog instruments are simple and direct reading devices that use a magnet (permanent or electromagnet) and a coil for their construction. The construction of a digital instrument, on the other hand, is more complex than that of an analog instrument due to the use of electronics and converter circuits for converting analog signals to digital signals. Although analog instruments can be used under any type of environmental conditions, digital instruments involve electronic devices that require proper environmental conditions in order to operate. However, analog instruments may produce considerable observational errors such as parallax errors. A digital instrument does not have moving parts, whereas an analog instrument does. Observing the output of analog instruments requires a greater amount of time. The accuracy of analog instruments is less than that of digital instruments, and analog instruments are less expensive. An analog instrument is normally larger than a digital instrument. In Figure 1.8, a digital caliper is shown for measuring a ring's outside diameter.

Null or ***deflection type:*** One of the operational modes of a measuring instrument is the null technique. An instrument of the null type determines the magnitude of the measured quantity based on a zero or null indication. The null type of instrument is equipped with either a manual or automatic balance that generates an equivalent opposing effect to counteract the physical effect of the quantity being measured. Thus, the equivalent null-causing effect provides a measure of the quantity. The equal arm balance scale is an example of a null instrument for measuring weight. Equal arm scales (see Figure 1.9) are one of the oldest types of scales in existence, dating back to around 1878 BC in ancient Egypt. In its

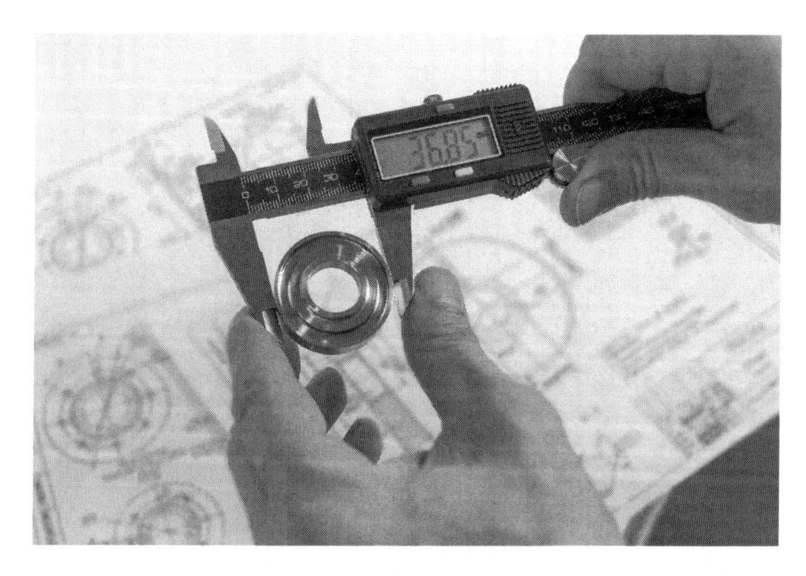

FIGURE 1.8 A digital caliper for measuring the ring's outside diameter.

(Courtesy: Shutterstock)

FIGURE 1.9 An equal arm scale.

(Courtesy: Shutterstock)

simplest form, the equal arm balance consists of two pans suspended on opposite sides of a lever. For it to be used, the object to be weighed (with an unknown weight) must be placed on one of the pans. Known weights are placed on the other pan until the indicator shows zero, indicating that the unknown object on one pan is equal to the known object on the other pan. A null instrument employs the null method for measuring the quantity. As a result of this technique, the instrument exerts an influence on the measured system in order to oppose the effect of the measurand. In order to achieve a null measurement, the influence and measurand must be balanced until their values are equal but opposite. In deflection-type instruments, the measured quantity produces physical effects that deflect or displace the moving system of the instrument. As an example, in a permanent magnet moving coil ammeter, the deflection of the moving point is directly proportional to the flow of current (the quantity under measurement). Compared with other operational modes of instrumentation, such as deflection instruments, a null instrument has certain inherent advantages. In order to minimize the interaction between the measuring system and the measurand, the null method balances the unknown input against a known standard input. Due to the fact that each input comes from a separate source, any measurement process influence on the measurand is minimal. Therefore, the measured system experiences very high input impedance, which minimizes loading errors. In particular, this technique is effective when the measurand is of a very small size.

This results in a high degree of accuracy for small input values and a low level of loading error when using the null operation. Null instruments have the disadvantage that iterative balancing operations require more time to perform than simple sensor input measurements. As a result, this technique may not offer the fastest measurement when high-speed measurements are required.

Smart or ***non-smart:*** Instruments that are smart, or intelligent, contain or incorporate a microprocessor. Instruments that are not smart, or dumb, do not contain a microprocessor. Users have been demanding better performance, easier maintenance, and more uptime from pneumatic instruments as they have evolved from simple pneumatics to sophisticated smart instruments. Smart instruments have met these demands and more, although their complexity has increased. When smart instruments are understood and deployed, the payoff is reduced complexity, improved performance, and reduced costs. As an example, suppose a process requires a minimum flow rate of 10 gpm (gallons per minute). The operator may set the flow to 11 gpm if he or she does not trust the accuracy of the flow measurement instrument, even if the excess flow increases costs both for raw materials and disposal. In the presence of a trusted and accurate smart flow instrument, the operator may feel comfortable reducing the flow to 10.1 gpm in order to increase efficiency and save money.

Over the past few decades, the definition of a smart instrument has evolved. In the past, dumb instruments were pneumatic devices controlled by single-loop controllers that operated at 3–15 psi. Generally, technicians recorded information manually using pens and paper as they made their rounds, often using gauges to display information. It was recommended to use a local chart recorder if data needed to be saved automatically and analyzed. Following this, a small degree of intelligence was added to instruments in the form of a 1–5, 4–20, or 10–50 mAdc output proportional to the process variable (PV). Data transmission provided the opportunity for remote measurement, display, and control once data could be transmitted remotely. Parallel development of control systems with central processing and I/O enabled more efficient means of capturing information produced by these 4–20 mA instruments, converting this information into engineering units, capturing this information centrally, and centrally acting on and recording it.

Indicating or ***recording:*** When an instrument type is indicating, the measuring quantities will only be indicated at a specific moment. It is not possible to store any previous values. For analog meters, the measuring values are indicated by a pointer and scale, while for digital circuits, they are displayed by a liquid crystal display (LCD) or light-emitting diode (LED) display. Voltmeters, ammeters, wattmeters, speedometers, and so on are examples of indicating instruments.

Consider a DC voltmeter with a range of 0–75 V. If the meter terminals are connected to 40-V supply, the pointer will indicate 40 V. However, if the supply is changed to 20 V, the pointer will immediately fall to 20 V. Consequently, the values are displayed immediately. Recording instruments are characterized by their names as being capable of recording measurements. The construction of these instruments is similar to that of indicating instruments, with the only difference being that in the case of an analog meter, the pointer and scale are replaced by a lightweight metal pencil and a piece of carbon paper. The power consumption recording meters at power plants are a good example of these instruments. Accordingly, a recording instrument is defined as an instrument that records the continuous variation of an electrical quantity over time. Normally, it is used in places where it is necessary to continuously monitor circuit conditions. For the purpose of future reference or computation, the record is kept. Voltmeters and thermoscopes are examples of recording instruments.

Absolute or ***secondary:*** The use of absolute instruments is very rare, except in standard laboratories for the purpose of standardizing. There is no need to calibrate or compare these instruments with any other instruments. These instruments give the magnitude of the quantity to be measured in terms of the deflection and the instrument constant. Secondary measuring instruments are the most widely used measuring instruments. It is necessary to calibrate these instruments prior to use. In order to measure the magnitude of an electrical quantity, the deflection of the instrument must be taken.

Mechanical or ***electrical*** or ***electronic***: Mechanical instruments are used to measure physical quantities. Because the instrument cannot respond to dynamic conditions, it is suitable only for measuring static and stable conditions. Examples of mechanical instruments include pressure gauges, speedometers, and water meters. Electrical instruments are used to measure electrical quantities such as current, voltage, and power. Electrical measuring instruments include ammeters, voltmeters, and wattmeters. An ammeter measures current in amps, a voltmeter measures voltage, and a wattmeter measures power. The output of an electrical instrument is faster than that of a mechanical instrument, which indicates that the output is more rapid. The reliability of electronic instruments is greater than that of other systems. Semiconductor devices are used in this device, and weak signals can also be detected.

1.5 INSTRUMENTATION AND MEASUREMENT SYSTEM CHARACTERISTICS

An understanding of the performance characteristics of a measurement system is essential to the selection process. A number of characteristics indicate the performance of an instrument, including accuracy, precision, resolution, and sensitivity. This allows users to select the most appropriate instrument for a particular measurement task.

The performance characteristics of measuring instruments can be divided into two categories:

1. A ***static characteristic*** is characterized by a slow change in the value of the measured variable. It is important to note that the static characteristics and parameters of measuring instruments provide only a description of the performance of the instruments with respect to steady-state input/output variables. There are several static characteristics and parameters that are intended to provide a quantitative description of the instrument's accuracy, and they are clearly described in the manufacturer's manuals and data sheets.
2. A ***dynamic characteristic*** is characterized by rapid changes in the value of the measured variable. Measurement instruments are characterized by dynamic characteristics that describe their behavior between the time they change value and the time at which they attain a steady value in response to that change in value. A dynamic characteristic value stated on an instrument data sheet only applies if the instrument is used in the specified environmental conditions.

1.5.1 Static Characteristics

1.5.1.1 Accuracy

The degree of accuracy (closeness) of a measurement is defined as the distance between the actual value and the measurement. The term is used in the manufacturer's specifications for measurement instruments and devices. The accuracy of an instrument refers to its ability to provide indications that are close to the true value of a variable being measured. The specifications of accuracy are actually expressed in terms of error (in other words, inaccuracies). It should be noted that when the accuracy of some measurement device is expressed as a percent error, we are able to estimate the error after measurement.

Example 1.1

Imagine a voltmeter with a range of 0 to 200 V and an accuracy of $x = 1\%$ of the range. There is a reading of 100 volts on the voltmeter. After measurement, what is the maximum amount of absolute error? In what range could the actual value be found?

Answer) Error = $1\% \times (200 - 0) = 2$ V

Consequently, the measurement value could differ by a maximum of 2 V from the actual value, which would result in an actual value of 100 V plus or minus 2 V at maximum or minimum, respectively, resulting in a value between 98 and 102 V.

1.5.1.2 Precision

The measure of consistency or repeatability is whether successive readings are not different from one another. As the name implies, precision refers to the capability of an instrument to show the same reading every time it is used (reproducibility). It is important to note that an instrument that is precise may not necessarily be accurate. As an example, an instrument frequently displays a measurement result with the same error. In this case, the instrument is precise but not accurate. A shooter aiming at the target can be used to demonstrate the difference between accuracy and precision (Figure 1.10). The precision of a shot is said to be high if all of the shots hit at the same point.

1.5.1.3 Sensitivity

This refers to the ability of an instrument to respond to small changes in the quantity that it measures. Alternatively, it can be defined as the ratio of change in output to change in input at a steady state. The resistance value of a platinum resistance thermometer changes as the temperature increases, for example. Therefore, Ohm/°C is the unit of sensitivity for this equipment.

In terms of a mathematical equation, sensitivity is defined as the ratio of change in output (response) compared to change in input at steady state.

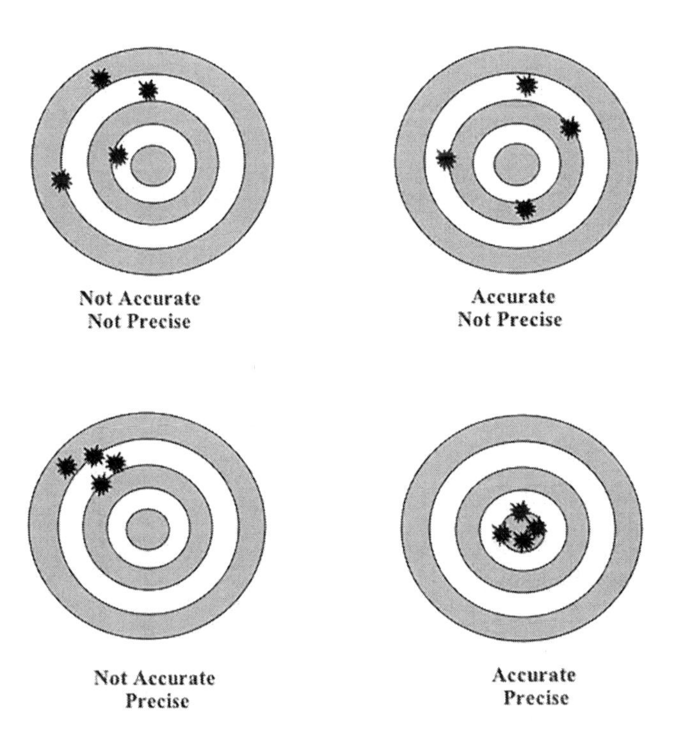

FIGURE 1.10 An illustration of accuracy vs precision.

$$Sensitivity\ (k) = \frac{\Delta\theta_o}{\Delta\theta_i} \tag{1.1}$$

Where

$\Delta\theta_o$ = Change in output
$\Delta\theta_i$ = Change in input

Example 1.2

To change the deflection of the galvanometer by 5 mm, it was necessary to change the arm of the bridge by 10 ohm. What is the value of sensitivity?

A. 2 ohm per millimeter
B. 0.5 millimeter per ohm
C. 0.5 ohm per millimeter
D. 2 millimeter per ohm

Answer) Option B is the correct answer.

1.5.1.4 Resolution

This is the smallest change in a measurement variable to which an instrument will respond. The resolution refers to the smallest change in input that can be measured.

Example 1.3

Identify the resolution of a voltmeter having a range readout scale of 50 divisions and a full-scale reading of 100 volts. If one tenth of a division can be read with certainty, then the voltmeter is considered to be high resolution.

Answer) 50 scale division = 100 V → One scale division = 100 ÷ 50 = 2 V
Resolution = 2 × 0.1 = 0.2 V

1.5.1.5 Reproducibility and Repeatability

The terms reproducibility and repeatability are commonly used interchangeably. It is important to note that repeatability and reproducibility are two components of precision in a measurement system. Repeatability refers to the closeness of measurement values between repeated measurements of the same thing, performed under similar conditions. Reproducibility refers to the closeness of measurements made of the same thing under different conditions. Essentially, reproducibility is the variation in readings when a different person measures the same part (or quantity) many times, using the same equipment (or different equipment), under the same conditions (or different conditions). If a person measures repeated readings of an object by micrometer as 15.02 mm, 15.03 mm, and 15.02 mm, then the person is competent and capable of repeating the readings. If three different individuals measure the same object by micrometer as 15.54 mm, 15.64 mm, and

15.49 mm, then the reproducibility of this measurement is 0.15 mm. As a result of reproducibility, the lab can demonstrate that its measurement results can be replicated under a variety of conditions. A lab that conducts a reproducibility test on every technician and finds that a technician's readings differ significantly from those of the others should provide them with appropriate training to improve their skills and increase their level of competence.

1.5.1.6 Range

An input range is defined by the minimum and maximum values of the variable (*Xmin* to *Xmax*), for example: –10°C to +150°C (for a temperature measurement device).

An output range of a measuring device can be defined by the minimum and maximum values of the output variable (*Ymin* to *Ymax*), for example, from 4 to +20 mA (for a measurement element with a current output).

1.5.1.7 Span

A measuring device's input span is determined by the difference between its maximum *Xmax* value and its minimum *Xmin* value: (*Xmax* – *Xmin*). When a measuring device has an input range of –10°C to +150°C, the input span is: +150°C – (–10°C) = 160°C.

A measuring device's output span is defined as the difference between the maximum *Ymax* and minimum *Ymin* values of the output variables: (*Ymax* – *Ymin*). For example, for a measuring device with an output range of 4 to +20 mA, the span would be: +20 – 4 mA = 16 mA.

1.5.1.8 Bias

A constant error that occurs during the measurement of an instrument. Calibration is usually used to correct this error.

1.5.1.9 Linearity

The maximum deviation from a linear relationship between input and output. Instruments must produce outputs that are linearly proportional to the quantities they measure. It is normally shown as a percentage of full scale (%fs). The graph in Figure 1.11 illustrates the output reading of an instrument based on a few input readings.

1.5.1.10 Threshold

An instrument's threshold defines the minimum value of input signal required to make a change or begin from zero. When the input is gradually increased from zero, this is the minimum value below which no output change can be detected. The output of a digital system is displayed in incremental digits. A threshold is defined as the minimum input signal required to produce at least one significant digit of output on a digital instrument. The difference between threshold and resolution of a measuring instrument can be understood in this manner. The threshold is defined as the smallest measurable input, while the resolution is defined as the

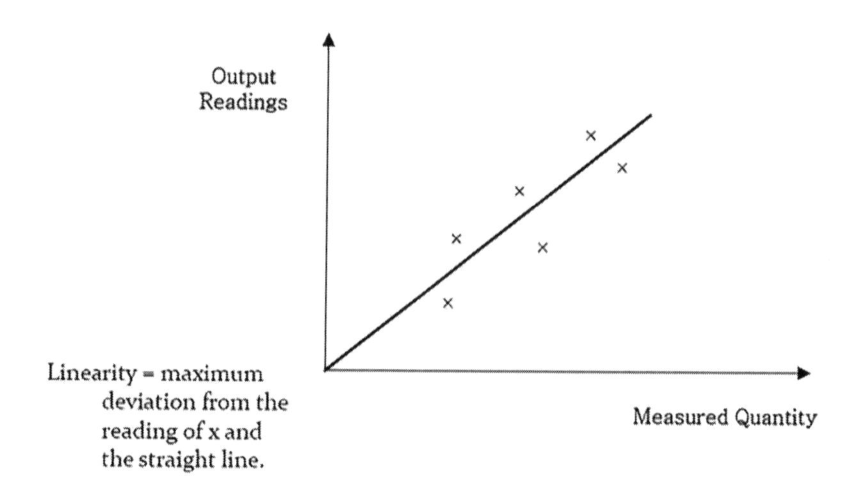

FIGURE 1.11 Linearity.

smallest measurable input change. It is possible to express both of these values in terms of an actual value or as a fraction or percentage of the full-scale value.

1.5.1.11 Expected Value

The desired value or most likely value is what is expected to measure.

1.5.1.12 Error

A variable's error (*e*) is the difference between its measured value and its true value.

1.5.1.13 Noise

An electronic circuit is characterized by noise, which is a random fluctuation in an electrical signal. Electronic devices generate a wide range of noise, as it can be caused by a number of different factors. In general, noise can be defined as an error or unwanted random disturbance of an information signal, introduced before or after the detector and decoder. A noise is the result of the accumulation of unwanted or disturbing energy from natural and sometimes human-made sources. It is often possible to express the quality of a signal quantitatively as a signal-to-noise ratio.

Any signal that does not convey any useful information may be considered noise. There is frequently a background on which a signal may be detected due to extraneous disturbances generated within the measuring system itself or coming from outside.

Noise comes from a variety of sources. Noise may originate at the primary sensing device, in a communication channel, or at other intermediate points. It is also possible for the indicating elements of the system to produce noise. The following are some of the most common sources of noise:

1. The presence of stray electrical and magnetic fields in the vicinity of the instruments produces extraneous signals that tend to distort the original signals. Shielding or relocation of the instrument components can minimize the effects of these stray fields.
2. Mechanical shocks and vibrations can also cause problems. The effects of these devices can be eliminated by using the appropriate mounting devices.

1.5.2 Dynamic Characteristics

Dynamic characteristics of a measuring instrument describe the behavior of the instrument between the time a measured quantity changes value and the time a steady value is obtained when the instrument output is adjusted to match the change in value. It should be noted that any values stated in an instrument data sheet regarding dynamic characteristics relate only to the instrument's performance under the specified environmental conditions. The following are a number of dynamic characteristics of an instrument: speed of response, measuring lag, fidelity, dynamic error, signal response, dynamic performance, and response time.

1.5.2.1 Speed of Response

It is defined as the rate at which a measurement system responds to changes in the quantity being measured. The speed of the instrument is indicated by this rate.

1.5.2.2 Measuring Lag

It refers to the delay in the response of a measurement system to changes in the measured quantity. A lag response delay is also known as a response lag. It is very important to reduce the time lag to the minimum in high-speed measurement systems.

1.5.2.3 Fidelity

The fidelity of a system is defined as its ability to reproduce its output in the same form as its input. Essentially, it is the degree to which a measurement system indicates changes in the measured quantity without introducing any dynamic errors. The system is said to have 100 percent fidelity if a linearly varying quantity is applied to it and if the output is also linearly varying. There should be 100% fidelity in a system.

1.5.2.4 Dynamic Error

Essentially, it is the difference between the true value of the quantity being measured, which changes over time, and the value indicated by the measurement system.

1.5.2.5 Signal Response

When an input signal is applied to an instrument, the signal response is its output response.

1.5.2.6 Dynamic Performance

This is a measure of how well a system responds to changing inputs.

1.5.2.7 Response Time

Response time refers to the amount of time it takes for an instrument or system to settle to its final stable position following the application of an input.

1.6 PRESSURE MEASUREMENT AND CONTROL

1.6.1 Introduction and Units

Most processes are affected by pressure, which is one of the most important variables. There are many other properties of systems that are affected by the operating pressure. For example, in an IC engine cylinder, the pressure changes continuously over time and determines the engine's output. The operating pressure has an impact on chemical reactions. In spite of the fact that pressure is not a primary quantity, it is derived from force and area. In turn, they are functions of mass, length, and time. The measurement of pressure is carried out in both static and moving fluid systems.

It is important to understand some of the pressure measuring quantities before we examine the various pressure measuring devices:

- 1 Pascal or 1 Pa = 1 N/m^2
- 1 atmosphere or 1 atm =760 mm mercury column = 1.013 × 105 Pa
- 1 mm mercury column = 1 Torr
- 1 Torr = 1.316 × 10^{-3} atm= 133.3 Pa
- 1 bar = 10^5 Pa

Pressure gauges are measured in relation to some absolute reference pressure, which is defined in a manner that is convenient to the measurement. There is a relationship between an absolute pressure, P_{abs}, and its corresponding gauge pressure, P_{gauge}, as follows:

$$P_{\text{gauge}} = P_{\text{abs}} - P_o$$

In this case, P_o is the reference pressure. The local absolute atmospheric pressure is commonly used as a reference pressure. It is a positive number when it comes to absolute pressure. Based on the difference between the measured pressure and the reference pressure, gauge pressure may be positive or negative. The atmospheric pressure at sea level is typically about 100 kPa, but it varies with altitude and weather conditions. In the event that the absolute pressure of a fluid remains constant, the gauge pressure of the same fluid will vary as the atmospheric pressure changes. A car, for example, will have a higher (gauge) tire pressure as it drives up a mountain due to the decrease in atmospheric pressure.

1.6.2 Pressure Gauges (Sensors)

Pressure sensors or piezometers (in hydraulic engineering) are devices used to measure the pressure of gases or liquids. One of the most frequently used instruments in any industrial facility is a pressure gauge, which measures the pressure within a system's media. In addition to temperature measurement, pressure measurement is an important measurement for operations in a wide range of applications—especially industrial applications—and it is vital to ensuring both the quality of a product and the safety of a facility and its personnel. As a measure of force required to stop a fluid from expanding, pressure is typically expressed as force per unit area. In most cases, pressure sensors act as transducers; they generate signals based on the applied pressure. A pressure sensor may also be referred to as a transducer, a transmitter, a sender, an indicator, a piezometer, or a manometer, among other names. There is a wide range of technology, design, performance, application suitability, and cost associated with pressure sensors. In some pressure sensors, the switch turns on or off when a certain pressure is reached.

The liquid column manometer is the oldest type of manometer and was invented in 1643 by Evangelista Torricelli. In 1661, Christiaan Huygens invented the U-tube. A U-tube manometer (see Figure 1.12) is one of the simplest gauges used to measure pressure. A U-tube with a uniform bore is oriented vertically, and the acceleration due to gravity is assumed to be known. The difference in height h between the levels of the manometer liquid in the two limbs of the U-tube is the quantity being measured. By doing so, the quantity to be measured, pressure, is converted to the length of a liquid column, its height. This is a pressure measurement of a system involving a fluid (liquid or gas) that is different from that of the

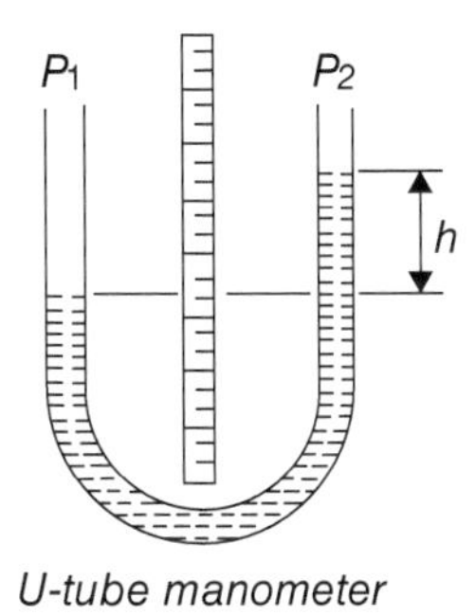

FIGURE 1.12 U-tube monomer.

manometer. The following figure illustrates the water levels in a U-tube where the left tube is connected to a higher pressure source than the right tube. For example, the left tube might be connected to a pressurized air duct, while the right tube is open to ambient air. Despite the fact that mercury is a common liquid used in manometers, other liquids are also used. Water is another common liquid.

The pressure difference measured by a vertical U-tube manometer can be calculated as

$$pd = \gamma h = \rho g h$$

where

pd = pressure (Pa, N/m^2, lb/ft^2)
$\gamma = \rho g$ = specific weight of liquid in the tube (kN/m^3, lb/ft^3)
ρ = U-tube liquid density (kg/m^3, lb/ft^3)
g = acceleration of gravity (9.81 m/s^2, 32.174 ft/s^2)
h = liquid height (m fluid column, ft fluid column)

Note: Typically, the specific weight of water in U-tube manometers is 9.81 kilograms per cubic meter, or 62.4 pounds per square foot.

Pressure gauges can use mechanical or electrical methods for measuring pressure. It is possible to sense pressure by using mechanical elements such as plates, shells, and tubes that are designed and constructed to deflect in response to pressure. As a result of this mechanism, pressure is converted into physical movement. When it comes to electrical methods of measuring pressure, it is then necessary to transduce this movement into an electrical output such as voltage. In addition, signal conditioning may be required, depending on the type of sensor and the application.

1.6.2.1 Mechanical Pressure Sensors

Normally, mechanical gauges use a dial or pointer to display the motion created by the sensing element.

1.6.2.1.1 Bourdon Tube Pressure Sensors

A Bourdon tube (see Figure 1.13) is a popular pressure gauge that can be used to measure the gauge pressure of liquids and gases. In the context of fluid media, gauge pressure refers to the pressure of the fluid media in relation to the atmospheric pressure. The Bourdon tube pressure gauge is a mechanical device. In order to function, it does not rely on an electrical signal. Fluid pressure is measured between 0.6 and 7,000 bar or 8 and 1,000 pounds per square inch (psi). Many industries use this pressure gauge since it can be used for low-pressure, vacuum, and high-pressure applications.

It is evident from the image that the Bourdon pressure gauge is attached to the pipeline of the system's inlet pipe. It is the socket or socket block that holds the inlet pipe in place, allowing pressure to flow into the stationary end of the tube. Between the fixed and moving ends of the elastic C-shaped tube, this pressure is

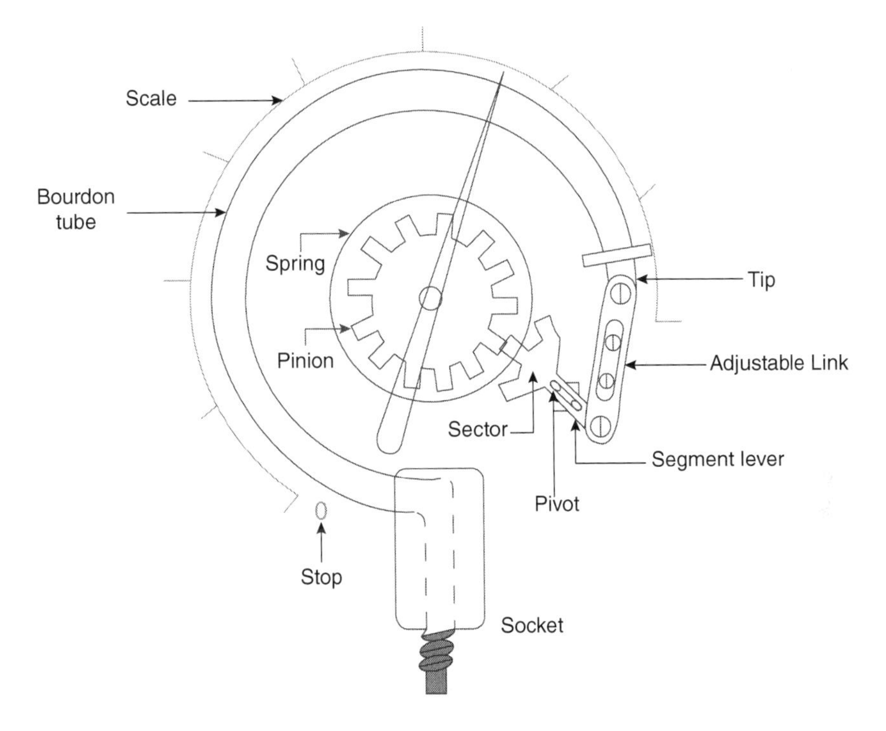

FIGURE 1.13 A Bourdon tube.

distributed throughout. In response to an increase in inlet pressure, the C-shaped tube straightens. This movement is connected to the sector gear by a pivot and pivot pin attached to the moving end of the tube. Consequently, every small change in inlet pressure results in an amplified motion, causing the indicator needle to deflect. In the event of an increase in inlet pressure, the indicator moves clockwise (from left to right) on a calibrated scale. The tube regains its helix shape when the pressure drops, and the indicator moves in an anticlockwise direction (from right to left). In summary, Bourdon tube gauges are based on the principle that a curved tube tends to straighten out under pressure. There is a pointer attached to the tube that indicates fluctuations in pressure on a calibrated dial.

1.6.2.2 Diaphragm Pressure Sensors

Diaphragm pressure gauges measure fluid pressure by measuring the deflection of a thin, flexible membrane. The diaphragm is a material that is elastic and will displace if pressure is applied to it. In this type of pressure gauge, the elastic property of the diaphragm will be used to determine the difference between the reference and unknown pressures. This device operates in the same manner as the Bourdon gauge. Diaphragm gauges differ from Bourdon gauges in that they have corrugated diaphragms rather than tubes. The small size of this type of pressure gauge makes it a popular choice. Diaphragm pressure gauges are best suited

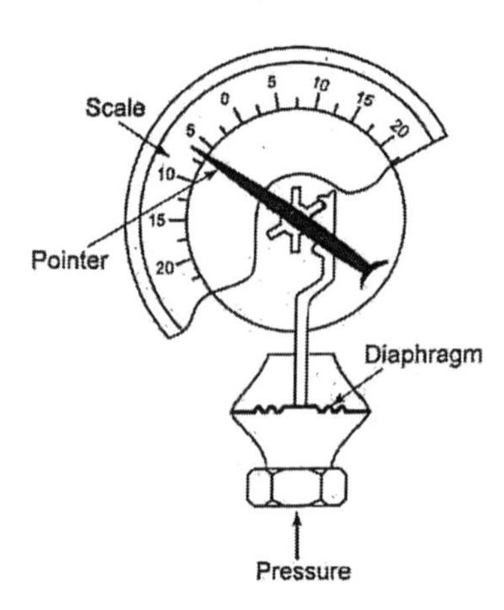

FIGURE 1.14 A diaphragm pressure gauge.

to low-pressure applications. The diaphragm isolates internal components from the media, making this gauge suitable for corrosive or contaminated liquids and gases. Figure 1.14 illustrates a diaphragm pressure gauge.

1.6.2.3 Bellows Pressure Sensors

Pressure gauges with bellows are composed of flexible metal bellows that expand or contract when pressure changes. Through the use of a connected mechanical or digital display, the expansion and contraction are converted into readable measurements. In addition to measuring absolute pressure, it is also capable of measuring differential pressure, positive and negative (vacuum) pressure. In many industrial and laboratory settings, these gauges are popular because of their high accuracy and durability. They are commonly used in applications with low to intermediate system pressures, such as process control, fluid power, and gas analysis. Bellows pressure gauges can be classified into two types: single bellows and dual bellows. A single-bellows pressure gauge (see Figure 1.15) measures the gauge or vacuum pressure of a system, whereas a two-bellows pressure gauge measures the absolute and differential pressures of the system.

1.6.2.4 Pressure Gauges with Capsule Sensors

This capsule element consists of two circular-shaped, convoluted membranes that are sealed tightly around their circumferences. Pressure is applied to the inside of the capsule, and the generated stroke movement is illustrated by a pointer as a measure of pressure. The use of pressure gauges with capsule elements is more suitable for gaseous media and relatively low pressures. As a matter of fact, the pressure that is to be measured is introduced into the capsule via an opening in the center of the first diaphragm. A transmission mechanism is connected to the center of the second diaphragm in order to transmit the deflection of the measuring element to the pointer. Both diaphragms will deform slightly when the pressure

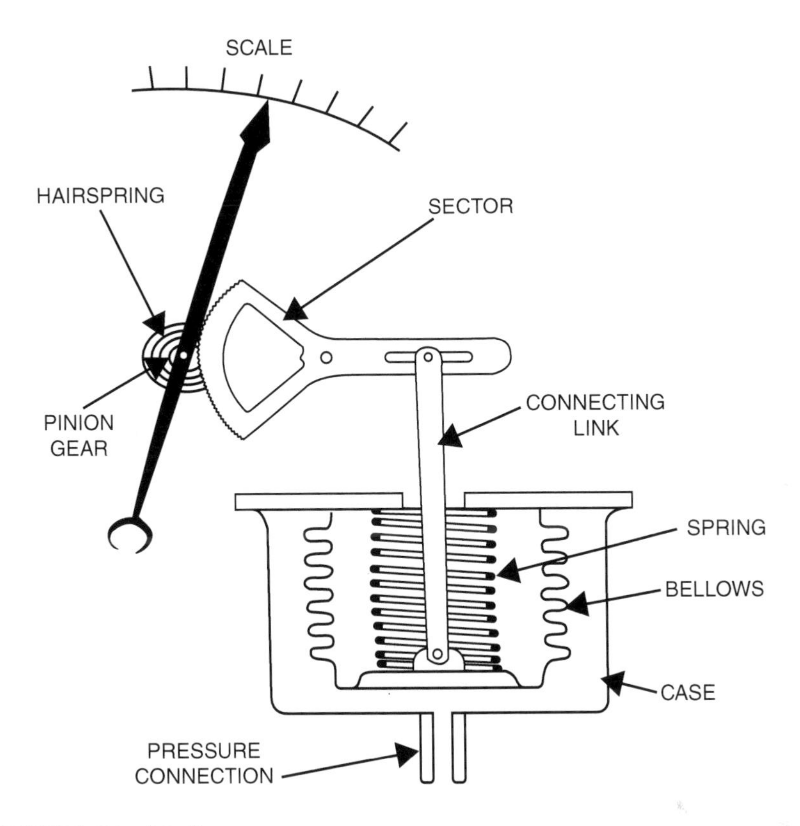

FIGURE 1.15 A bellows pressure gauge.

inside the capsule rises. In the presence of two diaphragms, the total deflection of the measuring element is twice as great. In the absence of two diaphragms, only very small deflections can be achieved due to the fast approach of the elastic limit. As a result, diaphragms are usually corrugated. As a result of these corrugations, the diaphragm can be deflected more than the elastic limit before it reaches the elastic limit. The magnitude of the deflection is also dependent on the diameter of the diaphragm. A diaphragm's deflection is proportional to its diameter multiplied by the fourth power. As the diameter of the diaphragm increases, the deflection increases rapidly. By doubling the diameter, the deflection increases by 16 times. Figure 1.16 illustrates a pressure gauge with capsule sensors.

1.6.2.5 Electrical Pressure Sensors

1.6.2.5.1 Piezoresistive Pressure Sensors

Piezoresistive strain gauges are one of the most common types of pressure sensors, derived from the ancient Greek word piezein (meaning to squeeze or press). In order to measure pressure, they use the change in electrical resistance of a material when it is stretched. A piezoresistive pressure sensor works by using a

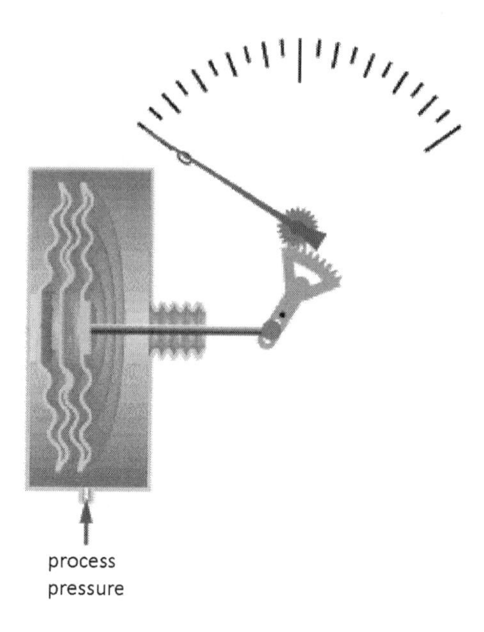

FIGURE 1.16 A pressure gauge with capsule sensors.

strain gauge made of a conductive material that changes its electrical resistance when stretched. Because of their simplicity and robustness, these sensors are suitable for a wide range of applications. Semiconductors and metals are typical piezoresistive materials that exhibit a strong piezoresistive effect. Typically, silicon is used in the manufacture of piezoresistive pressure cells, as well as in the production of electronic circuits. It is therefore also known as a sensor chip when sensors are made using this technology. In both high- and low-pressure applications, they can be used to measure absolute, gauge, relative, and differential pressures. Figure 1.17 shows a piezoresistive pressure sensor. As a matter of fact, strain gauge sensors (also referred to as strain gauge transducers) are capable of measuring this change in length caused by an external force and converting it into an electrical signal, which can then be converted into digital values, displayed, captured, and analyzed. When a strain gauge sensor is stretched or compressed, it experiences a change in resistance.

1.6.2.5.2 Capacitive Cell Pressure Sensors

Capacitive pressure sensors operate on the principle that if a differential pressure deforms the sensing diaphragm between two capacitor plates, there will be an imbalance in capacitance between the two plates. Using a capacitance bridge circuit, this imbalance is detected and converted into a direct current output of 4 to 20 mA. Metal, ceramic, and silicon diaphragms are commonly used in these technologies. Figure 1.18 shows a capacitive pressure sensor.

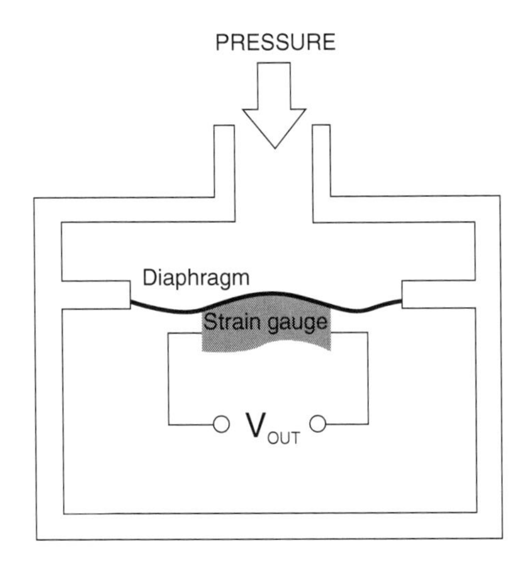

FIGURE 1.17 A piezoresistive pressure sensor.

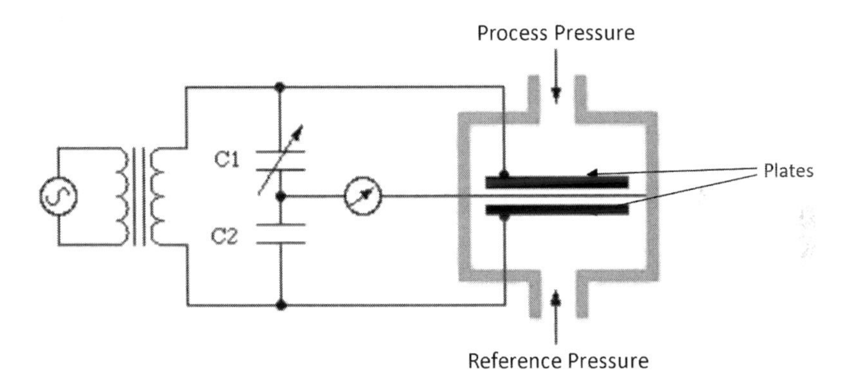

FIGURE 1.18 A capacitive pressure sensor.

1.6.2.5.3 Optical Pressure Sensors

By detecting changes in light, optical pressure sensors detect changes in pressure. For the simplest case, this can be achieved by a mechanical system that blocks the light as the pressure increases. In more advanced sensors, the measurement of phase difference can provide very accurate measurements of small changes in pressure. In an optical pressure sensor based on intensity, an increase in pressure will result in a progressive blocking of the light source. As a result, the sensor measures the change in light received. An example of this is shown subsequently in which the pressure moves a diaphragm and the attached opaque vane blocks more of the LED's light. As a result of the reduction in light intensity,

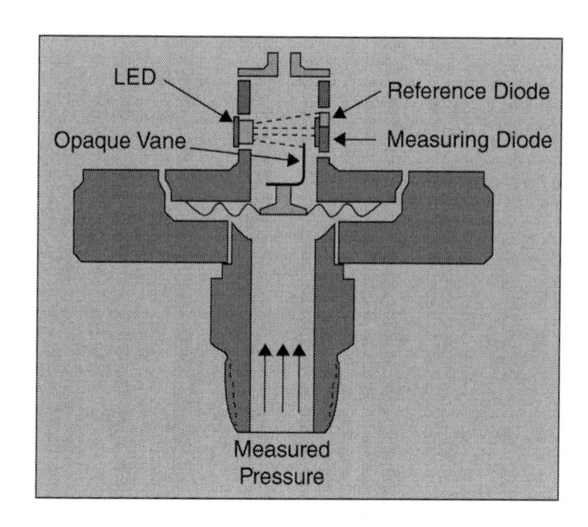

FIGURE 1.19 An optical pressure sensor.

a photodiode is able to detect the change in pressure directly. In other fiber-optic sensors, interference is used to measure changes in the path length and phase of light caused by changes in pressure. Accordingly, optical sensors work by modulating light traveling between a light source and a light detector. Figure 1.19 shows an optical pressure sensor.

1.6.2.5.4 Electromagnetic Pressure Sensors

There are many types of electromagnetic pressure sensors that use electromagnetic principles, including inductive pressure sensors, Hall pressure sensors, and eddy current pressure sensors.

1.6.2.5.4.1 Inductive Pressure Sensors An iron core and diaphragm are the main components of the variable reluctance pressure sensor (see Figure 1.20). As a result of the air gap between them, a magnetic circuit is formed. There is a change in the magnetic resistance when there is pressure in the air gap, resulting in a change in the air gap size. By applying a certain voltage to the iron core coil, the current will change as the air gap changes, thereby allowing the pressure to be measured. Different magnetic materials and permeabilities contribute to the working principles of the inductive pressure sensor. When pressure is applied to the diaphragm, the size of the air gap changes. It is important to note that the change in the air gap affects the change in the coil's inductance. In order to achieve the goal of measuring pressure, a corresponding signal output is generated.

The following paragraph describes the other possible type of inductive pressure gauge, as shown in Figure 1.21. When the distance between two magnetic devices is changed, the magnetic reluctance or magnetic resistance will change as well. By applying pressure to a portion of the magnetic circuit (movable core),

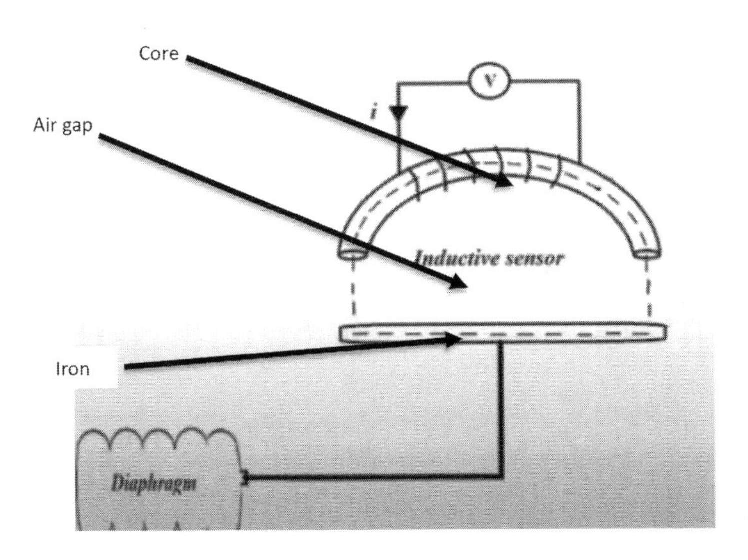

FIGURE 1.20 An inductive pressure sensor.

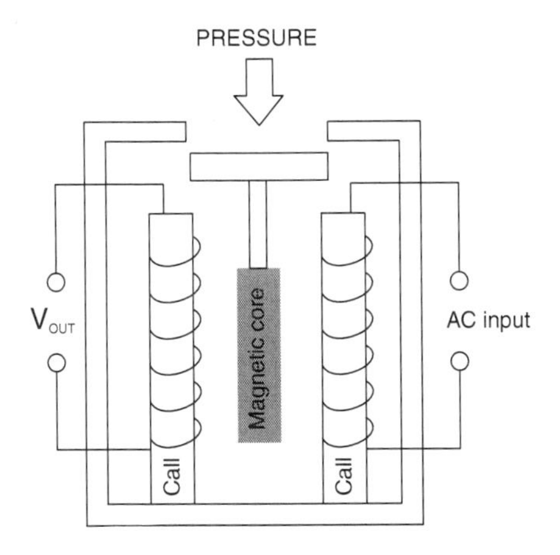

FIGURE 1.21 An inductive pressure sensor (an alternative design).

the reluctance between the coils is changed. The amplitude of the displacement in relation to the output voltage is used to calculate the pressure. The concept of magnetic reluctance, or magnetic resistance, is used in the analysis of magnetic circuits. In general, it is defined as the ratio between magnetomotive force (mmf) and magnetic flux. Magnetic flux is a measurement of the total magnetic field passing through a given area.

1.6.2.5.4.2 Hall Pressure Sensors Using the Hall effect, a Hall effect sensor (or simply Hall sensor) detects the presence and magnitude of magnetic fields and converts it to the voltage. How does the Hall effect work? In any material, as

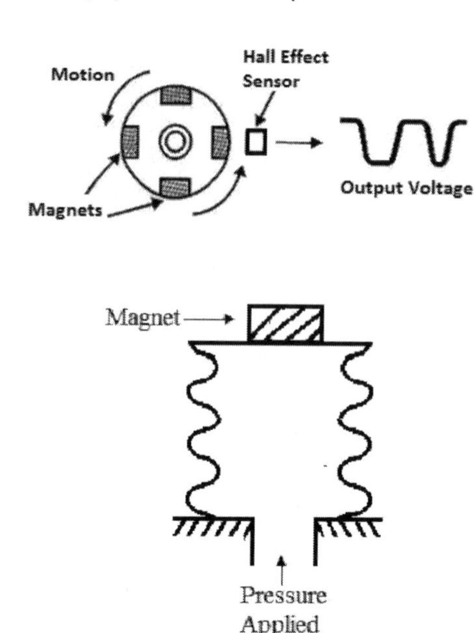

FIGURE 1.22 A Hall pressure sensor.

an electric current flow, electrons within the current naturally move in a straight line, creating their own magnetic field. The electrons will deviate into a curved path as they travel through an electrically charged material placed between the poles of a permanent magnet, rather than moving in a straight line. As a result of this new curved movement, more electrons are now present on one side of the electrically charged material. A potential difference (or voltage) will then appear across the material at right angles to the magnetic field due to both the permanent magnet and the electric current flowing through it. With the aid of a Hall sensor, which creates a magnetic field, and a magnet, the sensor converts the applied pressure into a differential Hall voltage output by deflection of a circular diaphragm with a simple rigid mechanical structure. A bellows is moved by the pressure, which in turn causes the magnet to move, which affects the output voltage created by the Hall effect sensor (see Figure 1.22).

1.6.2.5.4.3 Eddy Current Pressure Sensors As an electric current is passed through a coil in a probe assembly, eddy currents are created, which are fields of alternating magnetic fields (see Figure 1.23). An eddy current sensor is generally referred to as a gap sensor. An eddy current is generated on the surface of a measuring object in the presence of a magnetic field, thereby altering the impedance of the sensor coil (effective resistance) and voltage. A correlation can be drawn between the voltage and the distance of the object being measured and, finally, the pressure being applied.

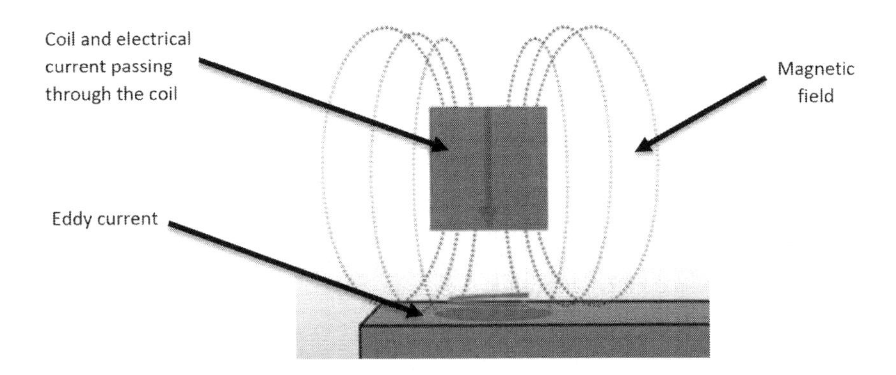

FIGURE 1.23 Eddy current pressure sensors.

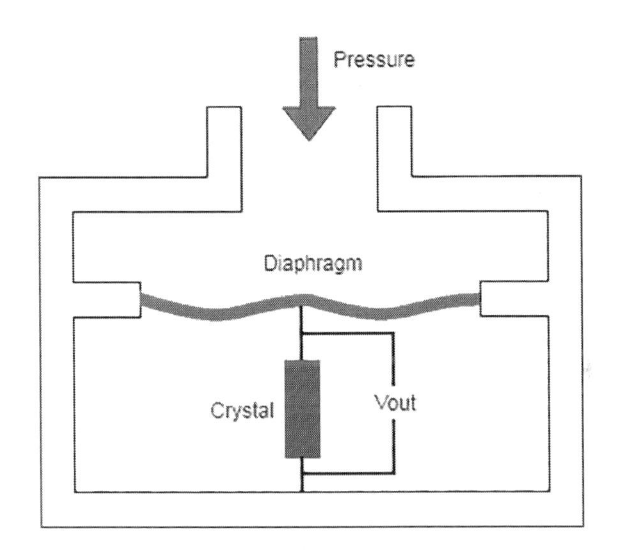

FIGURE 1.24 A piezoelectric pressure sensor.

1.6.2.5.5 Piezoelectric Pressure Sensors

Piezoelectric pressure sensors are based on a crystal that produces an electrical charge in proportion to the applied pressure. Approximately 40 crystalline materials are capable of generating electric charges when strained. This is commonly referred to as the piezoelectric effect. The system measures the strain on the sensing mechanism as a result of pressure by using the piezoelectric effect in certain materials, such as quartz. A common application of this technology is the measurement of highly dynamic pressures. With piezoelectric sensors, static pressures cannot be measured since the basic principle is dynamic. As the name implies, static pressure is the pressure a fluid would exert on its surroundings if it were not moving. Dynamic pressure, also called velocity pressure, is caused by the fluid's velocity. An example of a piezoelectric pressure sensor can be seen in Figure 1.24.

1.6.2.5.6 Potentiometric Pressure Sensors

In order to obtain an electronic output from a mechanical pressure gauge, a potentiometric pressure sensor (see Figure 1.25) is used. This device consists of a precision potentiometer, whose wiper arm is mechanically connected to a Bourdon or bellows element. Using a Wheatstone bridge circuit, the movement of the wiper arm across the potentiometer converts mechanically detected sensor deflection into a resistance measurement. This type of measurement is subject to unavoidable errors due to the mechanical nature of the linkages connecting the wiper arm to the Bourdon tube, bellows, or diaphragm element. As a result of the different thermal expansion coefficients of the metallic components of the system, temperature effects cause additional errors. The components and the contacts may also wear mechanically, resulting in errors.

1.6.2.5.7 Force-Balancing Pressure Sensors

The self-balancing system is an important traditional technology for all types of continuous measurement. “Self-balancing” systems constantly balance adjustable quantities against sensed quantities, with the adjusted quantities becoming indicators of the sensed quantities once balance is achieved. To measure mass, laboratories commonly use manual-balance systems. A sensed quantity is the unknown mass, whereas an adjustable quantity is the known mass. Human lab technicians apply as many masses to the left-hand side of the scale as necessary in order to achieve balance and then count up the total amount of those masses to determine the quantity of the unknown mass. By acting on the surface area of a sensing element such as a diaphragm or a bellows, pressure can easily be converted into force in pressure instruments. An instrument with force-balanced pressure may be created by producing a balancing force to cancel out the force exerted by the process pressure.

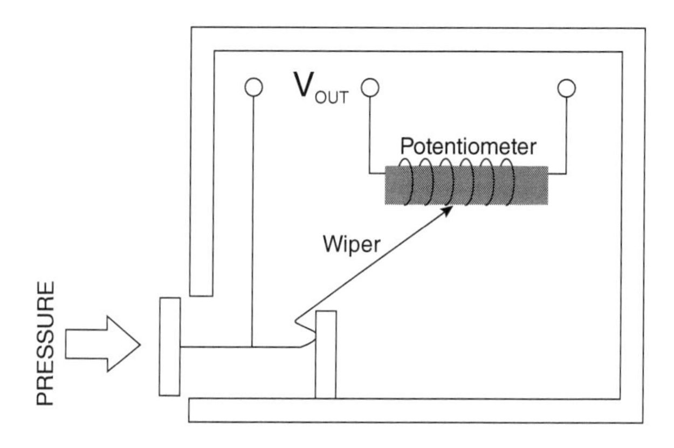

FIGURE 1.25 A potentiometer pressure sensor.

1.7 TEMPERATURE MEASUREMENT AND CONTROL

The measurement of temperature is one of the most common and most important measurements in industrial processes. There are many applications and needs for temperature measurement in today's industrial environment. As a result of the wide variety of needs in the process controls industry, a large number of sensors and devices have been developed to meet this demand. Temperature is a critical and widely measured variable for most mechanical engineers. It is necessary to monitor or control the temperature in many processes. This can range from simple monitoring of the water temperature of an engine or load device to monitoring the temperatures of fluids in processes or process support applications or of solid objects such as metal plates, bearings, and shafts in a piece of machinery. Today, a wide variety of temperature measurement probes are available depending on the type of measurement required, the accuracy needed, and the purpose of measurement.

A temperature can be thought of as a measure of the energy inherent in a body, which is the result of the random movement of its atoms or molecules. Temperature is a state variable that, along with mass, heat capacity, and other quantities, describes the energy content of a body or system. As a result, temperature can be measured directly in terms of energy. In spite of this, the tradition of specifying temperature in degrees had already been introduced much earlier and was well established in physics, so it was not reasonable to change this practice. Currently, three main temperature scales are commonly used in the world to measure temperature: Fahrenheit (°F), Celsius (°C), and Kelvin (K). There are a number of different scales that use different divisions based on different points of reference. In the 16th and 17th centuries, the first thermal instruments were developed. The melting point of ice and the boiling point of water were proposed as the standard in 1665.

1.7.1 Temperature Measurement Classifications

In general, temperature measurement can be categorized into three general categories: (1) *thermometers*, (2) *probes*, and (3) *noncontact* temperature measurement instruments.

1.7.1.1 Thermometers

Thermometers measure temperature or temperature gradients (the degree to which an object is hot or cold). There are two important components of a thermometer: (1) a temperature sensor (e.g. a mercury-in-glass thermometer bulb or other liquid) whose function changes as the temperature changes and (2) a means to convert this change into a numerical value (for example, the visible scale on mercury-in-glass thermometers or the digital readout on infrared models). Technological and industrial applications of thermometers include monitoring processes, meteorology, medicine, and scientific research.

Among the group of instruments, thermometers are the oldest. There was a need to measure and quantify the temperature of something around the year 150 AD. It was not until the development of science in the 1500s that thermometry evolved into a science. The first thermometer was described in *Natural Magic* in the 16th century and was an air thermoscope. A tube with a bulb and stem was partially filled with alcohol by Ferdinand II de' Medici, Grand Duke of Tuscany, in 1654. This was the first thermometer that was dependent on the expansion and contraction of a liquid, independent of barometric pressure. Various variations of this concept were developed, each with its own characteristics due to the lack of a standard scale. In 1665, Christian Huygens proposed using the melting point of ice and the boiling point of water as standards. In Copenhagen, Danish astronomer Ole Reiner used these upper and lower limits for a thermometer to record the weather. The effectiveness of these parameters at different geographical latitudes was still uncertain. Carlo Renaldini proposed a universal scale based on the limits of ice and boiling water in 1694. Between melting ice and body temperature, Isaac Newton proposed a scale of 12°C in 1701. The following are some examples of thermometer types.

1.7.1.1.1 Glass-Type Thermometers

In addition to being a small, compact instrument of durable steel construction, the glass tube industrial thermometer is also known as a liquid in glass thermometer. Most commonly, liquid in glass thermometers is used to measure temperature since they are inexpensive to produce and easy to use. This thermometer combines the advantages of easy readability, small size, and adaptability to a variety of applications that require an economical thermometer. Pipelines of all types, commercial building applications (heating and cooling), process piping, tanks, boilers, and so on are all common applications for liquid-in-glass industrial thermometers.

From the bulb, there is a very thin opening, also known as a bore, which extends down the center of the tube. Typically, the bulb is filled with mercury or red-colored alcohol, and when the temperature increases, it expands and rises into the tube, and when the temperature decreases, it contracts and moves downward. In order to measure temperature, the apparent thermal expansion of a liquid is used. The background of the glass tube is covered with white paint, while the front of the tube forms a magnifying glass that makes it easier to determine the temperature of the liquid column. In Figure 1.26, a glass thermometer filled with red alcohol is shown.

Glass thermometers are still used for some highly precise measurements. The properties of fluids, in particular mercury, are well known, so the only limitation to accuracy and resolution is the ability to manufacture glass tubes with precision bores. Due to the hazards associated with spilled mercury, many thermometers today use fluids other than mercury. Other fluids are used in these newer devices that have been engineered to exhibit specific expansion rates. The disadvantage of these fluids is that they do not typically have the thermal properties of mercury, which allows them to be used at high temperatures. The glass thermometer has

FIGURE 1.26 A glass-type thermometer.

(Courtesy: Shutterstock)

the major disadvantage of having a limited pressure capacity. Additionally, the accuracy of the thermometer was compromised when the glass bulb was inserted into a pressurized fluid or chamber.

1.7.1.1.2 Bimetal Thermometers

A bimetallic strip is used as a mechanical temperature sensor. Temperature is converted into mechanical displacement by this device. The bimetal thermometer is a thermometer based on the principle that metals expand differently when temperatures change. There are two different metal strips in a bimetal thermometer, each with a different coefficient of thermal expansion. Using a bimetallic strip, the temperature of the media is converted into mechanical displacement. Bimetallic strips contain two different metals with different coefficients of thermal expansion. The property of thermal expansion is the ability of a metal to change its shape or volume as a result of a change in temperature. By fusing or riveting the metal strips together along their length, the strips are connected along their length. On one end, the strips are fixed, while on the other end, they are free to move. Steel and copper are the two most commonly used metals; however, steel and brass may also be used. Because these metals exhibit different degrees of thermal expansion, as illustrated in Figure 1.27, their lengths change at different rates for the same temperature. Due to this property, when the temperature changes, one side of the metal strip expands, while the other does not, resulting in bending. During times of rising temperatures, the strip will turn towards the metal with a lower temperature coefficient. Strips bend in the direction of metals with higher temperature coefficients when the temperature decreases. Temperature variation can be determined by the deflection of the strip. The temperature of

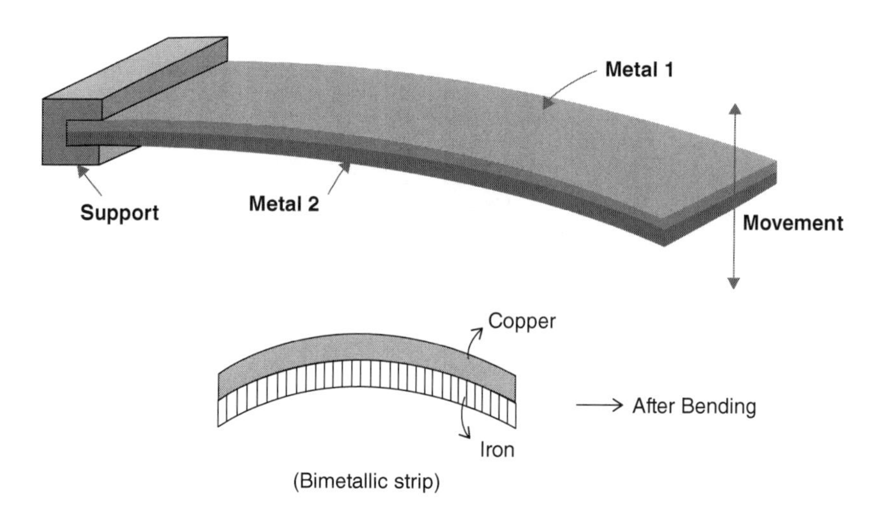

FIGURE 1.27 Bi-metallic bending in a thermometer.

the media is determined by this bending motion, which is connected to the dial on the thermometer. There are several applications for bimetallic thermometers, including residential devices such as air conditioners and ovens, as well as industrial devices such as heaters, hot wires, and refineries. A thermometer is a simple, reliable, and cost-effective means of measuring temperature.

1.7.1.2 Probes

A temperature probe with electrical thermometers is a device that consists of one or more temperature sensors and an armor tailored specifically for the intended use. Thermocouples, resistance temperature detectors (RTDs), and thermistors are three types of temperature probes that are discussed in more detail in the following.

1.7.1.2.1 Thermocouples

A magnetic field is observed when different metals are joined at the ends by a temperature difference, according to the German physicist Thomas Johann Seebeck in 1821. While Seebeck referred to this phenomenon as thermomagnetism at the time, he later demonstrated that the magnetic field he observed was the result of thermoelectric currents. When used in practice, the voltage generated at the junction of two different types of wire is of interest since this can be used to measure the temperature at very high and low temperatures. Depending on the type of wire used and the temperature, the magnitude of the voltage will vary. Thermocouples are devices capable of converting heat energy into electrical energy. Following the development of the thermometer, the temperature probe was the next step in the evolution of temperature measurement. A thermocouple composed of platinum and palladium was first used in 1826 by an inventor named Becquerel. Prior to this time, all temperature measurements were performed with liquid- or gas-filled thermometers. Upon the invention of the thermocouple, a whole new wave of development began, culminating in what we know today as practical thermometry. Thermocouples are sensors that measure temperature. There are two different types of metals that are joined at one end. When the junction between two metals is heated or cooled, a voltage is generated that can be correlated with the temperature (see Figure 1.28). Thermocouples are simple, robust, and cost-effective temperature sensors that are widely used in temperature measurement applications. In boilers, water heaters, ovens, and aircraft engines, to name just a few, they are used in a variety of applications up to approximately +2500°C. A total of 13 types of thermocouples are commonly used as "standard" thermocouples. A total of eight types have been assigned internationally recognized letter type designations. A thermocouple type can be matched to a specific application based on its characteristics. K and N types are generally preferred by industry due to their suitability for high temperatures, whereas T types are preferred by others due to their sensitivity, low cost, and ease of use. As explained, most thermocouples are of type K, with a measurement range of–200°C to +1250°C, consisting of chrome and Alumel (trademarked nickel alloys containing chromium and aluminum, manganese, and silicon, respectively). The color coding of thermocouple wire is important. At least seven different standards exist. There are some inconsistencies between the standards, which appear to have been designed to cause confusion. As an example, the color red is always used for the negative lead in the US standard, while in the German and Japanese standards, it is always used for the positive lead. There are no British, French, or international standards that use red at all.

As summarized in this paragraph, thermocouples have both advantages and disadvantages. These devices have the advantage of being able to be used for a wide range of temperature applications. It is possible to use thermocouples to measure most practical temperature ranges, from cryogenics to jet engine exhaust. Thermocouples can measure temperatures ranging from –200°C to +2500°C, depending on the metal wires used. In addition, thermocouples are

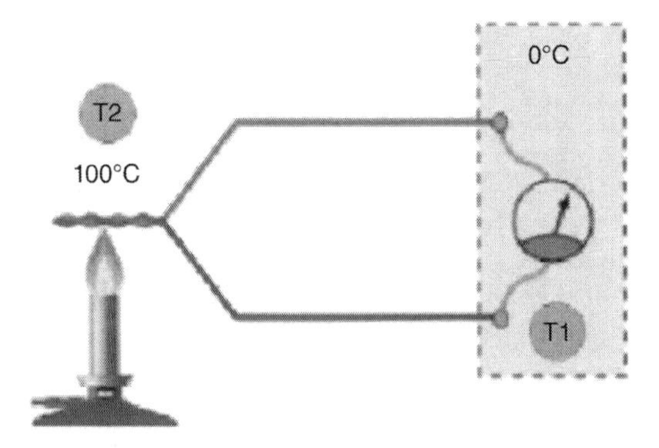

FIGURE 1.28 A thermocouple.

robust devices that are immune to shock and vibration and are suitable for use in hazardous environments. Due to their small size and low thermal capacity, thermocouples respond rapidly to temperature changes, especially if the sensing junction is exposed to the outside environment. Within a few hundred milliseconds, they can respond to rapidly changing temperatures. Finally, thermocouples are intrinsically safe since they do not require excitation power and do not self-heat. The disadvantages of thermocouples include complex signal conditioning, inaccuracy, corrosion susceptibility, and noise. In order to convert the thermocouple voltage into usable temperature readings, substantial signal conditioning is required. Additionally, thermocouples have inherent inaccuracies due to their metallurgical properties. A thermocouple measurement is only as accurate as the reference junction temperature, usually within 1°C to 2°C. Because thermocouples are made up of two dissimilar metals, corrosion over time may deteriorate their accuracy in certain environments. Consequently, they may need protection; care and maintenance are essential. Noise from stray electrical and magnetic fields can be a problem when measuring microvolt-level signal changes.

1.7.1.2.2 Resistant Temperature Detectors

The resistance thermometer, also known as the resistance temperature detector (RTD), is a device that measures the temperature. Resistance temperature detectors are based on the predictable and repeatable phenomenon of electrical resistance changing as a function of temperature. There is a similar temperature coefficient for all pure metals—0.003 to 0.007 ohms/ohm/°C. A number of metals are commonly used for temperature sensing, including platinum, nickel, copper, and molybdenum. The resistance-temperature characteristics of certain semiconductor and ceramic materials can be used for temperature sensing, but these sensors are not generally classified as RTDs. In 1871, Sir William Siemens proposed platinum as an element for a resistance temperature detector: it is a noble metal and has the most stable resistance–temperature relationship over a wide range of temperatures. The resistance of nickel elements is limited by temperature because the resistance changes non-linearly with temperature at temperatures above 300°C (572°F). The resistance of copper is linear with temperature; however, copper oxidizes at moderate temperatures and cannot be used over 150°C (302°F).

The manufacture of RTDs can be accomplished in two ways: by using wire or by using film (see Figure 1.29). A wire RTD consists of a coil of fine wire that is stretched in a ceramic tube that supports and protects the wire. A glaze may be used to bond the wire to the ceramic. It is generally the wire types that are more accurate, due to the tighter control over the purity of the metal and fewer strain-related errors. Additionally, they are more expensive. Film RTDs are manufactured by silk-screening or vacuum sputtering a thin metal film onto a ceramic or glass substrate. Compared to wire sensors, film sensors are less accurate, but they are relatively inexpensive, are available in small sizes, and are more robust.

The purpose of this paragraph is to compare RTDs with thermocouples. It is best to use thermocouples when working at high temperatures. New manufacturing

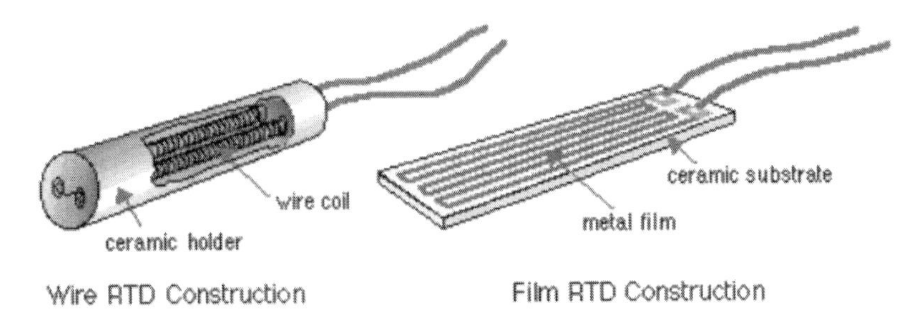

FIGURE 1.29 Wire RTD and film RTD.

techniques have improved the measurement range of RTD probes, but more than 90% of RTDs are designed for temperatures below 400°C. RTDs are generally more expensive than thermocouples. RTDs typically cost two to three times more than thermocouples with the same temperature range and style. In spite of the fact that both types of sensors respond quickly to temperature changes, thermocouples respond more rapidly. Thermocouples are generally less accurate than RTDs. For a long period of time, RTD probe readings remain stable and repeatable. As a result of chemical changes in the sensor (such as oxidation), thermocouple readings tend to drift.

1.7.1.2.3 Thermistors

In electrical engineering, a thermistor is a resistor whose value changes as the temperature changes. There are two pieces in the word thermistor, the first being thermal, which means heat, and the second being resistance. Thermistors can be categorized into two main categories. The first is negative temperature coefficient (NTC), and the second is positive temperature coefficient (PTC). A negative temperature coefficient thermistor is a resistor whose resistance value decreases as the temperature increases. Thermistors with positive temperature coefficients have resistance values that increase as the temperature increases.

Typically, a thermistor is composed of sintered metal oxides embedded in ceramic matrixes that change electrical resistance as a function of temperature. In addition to being sensitive, they are also highly nonlinear. Their high sensitivity, reliability, ruggedness, and ease of use have made them popular in research applications, but they are less commonly used in industrial settings. The most desirable materials are those with predictable values of change. A thermistor was originally made from loops of resistance wire, but modern thermistors are made of sintered semiconductor materials that can change resistance with a small change in temperature. The main difference between an RTD and a thermistor is that RTDs are constructed from different metals, whereas thermistors are made from ceramic materials. As a result of the material used in the manufacture of the thermistor, there will be a variation in the resistance. In its most common form, a thermistor consists of a beaded element attached to two wires.

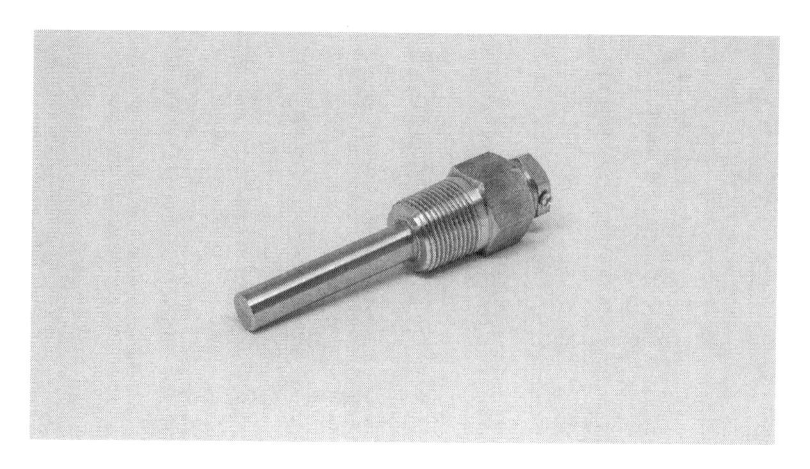

FIGURE 1.30 A thermowell.

(Courtesy: Shutterstock)

1.7.1.2.4 Thermowells

In industrial applications, thermowells are cylindrical pressure-tight fittings used to protect temperature sensors such as thermocouples, thermistors, and bimetal thermometers that are inserted into pipes or vessels. Basically, a thermowell is a tube that is closed at one end and mounted in a process stream. During the process of sensing, the thermowell acts as a barrier between the sensing element and the process medium. As a result, it protects the sensing element against corrosive process media and fluid pressures and velocities. A thermowell in stainless steel is illustrated in Figure 1.30.

1.7.1.2.5 Semiconductor Probes

The other type of probe is a semiconductor probe. As with resistance probes, they require a current (or voltage) supply to generate a reading. This is the extent of the similarity. A semiconductor probe consists of semiconductor material containing a number of active circuits. It consists primarily of a temperature-dependent resistance device, which converts the change in resistance into a change in current. As an example, the controlled current output is equal to 1 amp per degree of temperature in Kelvin.

1.7.1.3 Noncontact (Remote) Temperature-Measuring Devices

With noncontact temperature measurement, the surface temperature can be determined without any physical contact between the measurement object and the temperature sensor. A wide range of noncontact temperature sensors are available, primarily optical devices. They all rely on some form of radiative heat transfer measurement. Heat is radiated by all things in general. Heat can be detected from a device as a form of radiation. A few millimeters away as well as millions of light years away can be measured by measuring this radiation in order to determine the temperature. Noncontact temperature measurement offers the following advantages:

1. It is fast (within a few milliseconds). A significant amount of time is saved, allowing for more measurements and data accumulation.
2. It facilitates the measurement of moving targets (conveyor processes).
3. Measurements can be taken of hazardous or physically inaccessible objects.
4. Measurements of high temperatures (up to 3,000°C) are not problematic.
5. The target is not interfered with, and no energy is lost. As compared to measurements made with contact thermometers, measurements made with infrared thermometers are extremely accurate when measuring materials with poor heat conductivity, such as plastic or wood.
6. The surface of the object is not affected by contamination or mechanical forces; therefore, there is no wear. It is possible to measure soft surfaces as well as lacquered surfaces, for example.

1.7.1.3.1 Infrared Noncontact Temperature Sensors

Infrared radiation (also known as infrared light or IR) consists of electromagnetic radiation (EMR) with wavelengths longer than those of visible light and shorter than those of radio waves. As a result, it is invisible to the human eye. The infrared temperature sensor measures the surface temperature without being in contact with it. Infrared temperature sensors measure the infrared energy emitted from an object and correlate it with the surface temperature. Apart from their use as noncontact temperature sensors, some models are also capable of measuring the temperature of very small objects due to their very fast response times. Noncontact thermometers or temperature guns, which describe the device's ability to measure temperature from a distance, are sometimes referred to as laser thermometers since lasers are used to help aim the thermometer. An infrared temperature sensor works by emitting an infrared beam of energy that is focused by a lens onto a surface to detect the surface temperature or the temperature of an object. This reflected beam is received by a sensor, which converts it into an electrical signal that can be displayed in units of temperature. In order to provide accurate temperature readings, the sensor accounts for ambient temperature variations. As shown in Figure 1.31, an infrared temperature sensor is used by an operator to measure the temperature of the surface of a pipe carrying water.

An IR temperature measure device must be able to see the target optically (infrared-optically). Measurements are less accurate when there is a high level of dust or smoke present. It is possible only to measure the surface of concrete barriers, such as a closed metallic reaction vessel, but it is not possible to measure the inside of the vessel. The optics of the sensor must be protected from dust and condensing liquids.

FIGURE 1.31 The use of an infrared temperature sensor to measure the temperature of water flowing through a pipe.

(Courtesy: Shutterstock)

1.7.1.3.2 Pyrometers

Pyrometers are types of remote-sensing thermometers that measure the temperature of distant objects. Pyrometers have existed in various forms throughout history. It is a device that from a distance can be used to determine the temperature of a surface based on the amount of thermal radiation that an object emits, also referred to as pyrometry or radiometry. The term radiometry refers to a set of techniques used to measure electromagnetic radiation, including visible light. The basic principle of the pyrometer is that it measures the object's temperature by sensing the heat/radiation that is emitted by the object without making contact with it. In contrast to a resistance temperature detector or thermocouple, the main advantage of this device is that there is no direct contact between the pyrometer and the object whose temperature needs to be measured.

1.8 LEVEL MEASUREMENT AND CONTROL

The level refers to the height at which a liquid or bulk material is filled, for example, in a tank or reservoir. Surfaces are generally measured with reference to a reference plane, usually the tank bottom. If the product's surface is not flat (for example, if it contains foam, waves, turbulence, or coarse-grained bulk material), level is defined as the average height within a defined area. Level gauges are devices that show the level of fluids in fields. There are a variety of industrial processes that use these instruments to monitor fluid levels in drums, tanks, pressure vessels, and other similar applications. Choosing the type of level gauge depends on the application. The process industry, for instance, uses tubular level indicators or magnetic level gauges to provide a better visual indication of liquid levels. It is necessary to use radar-type indicators or ultrasonic-type indicators when measuring levels noncontact. The level gauge is a direct method of measuring the level. As a level gauge, a measuring chamber is connected in parallel to the vessel being monitored so that the level can be directly indicated visually or using a magnetic indicator or transducer. Level gauges are often used in process vessels as well as storage vessels.

1.8.1 Level Gauge Classifications

It is possible to find the following types of level gauges depending on the application and level measuring technique: tubular glass level gauges, transparent level gauges, reflex level gauges, magnetic level gauges, bicolor level gauges, radar level gauges, ultrasonic level gauges, float level gauges, and hydrostatic head devices.

1.8.1.1 Tubular Glass Level Gauges

A tubular level indicator or a tubular glass level indicator is the simplest form of level indication. Designed for low-pressure, non-toxic applications, tubular level gauges are used to measure and report the level of process fluid in tanks and vessels by directly reading the level. Tubular glass gauges consist of a transparent glass tube, seals, end blocks, valves, nozzles, vent, drains, and guard rods to protect the glass. In order to indicate the level, it must be positioned parallel to the vessel along the elevation where the level is to be indicated and be fitted with suitable fittings for retaining pressure and sealing the ends of the sight tube. The construction of this device, however, is not well suited for use with hazardous process fluids. Keeping personnel and equipment safe is an important consideration in the selection of gauges. There may be difficulties when high-temperature, high-pressure, corrosive, or other dangerous fluids or steam are used in the process vessel, storage tank, pipeline, and so on. There is a potential for a hazardous condition if the sight glass tube suffers a fracture of the glass or a leak at the seals. Transparent level gauges are always fitted with two transparent glass plates between which the fluid is contained. The level of liquid inside the glass tube is the same as the level of liquid in the tank. The glass tube level indicator is designed using the same principle as the connector. Upon forming a liquid path between the liquid in the tank and the glass tube, the liquid will flow towards the glass tube under the influence of air pressure until the two sides remain level. Figure 1.32 shows a tubular glass level gauge.

1.8.1.2 Transparent Level Gauges

It is highly beneficial to use transparent level indicators in chemical industries and petrochemical fertilizers. Due to the high pressure and high temperature at which the fluid is stored, a transparent level gauge is very useful for measuring the fluid level. A transparent level gauge differs from a glass tube level gauge in that it consists of two plates of transparent glass, between which the fluid is contained. It should be noted that a single plate of transparent glass is possible. The transparent

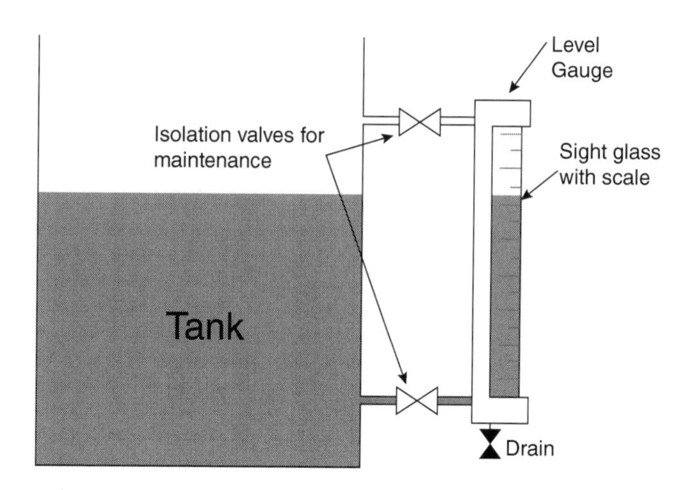

FIGURE 1.32 A tubular glass level gauge.

measuring chamber for the level gauge holds the fluid and accepts the glass and covers, which are secured by bolts. It is possible to view the fluid level through one or more vision slots that are machined into the chamber. In the vision slot, tie bars may be left (not machined out) to provide additional strength. In order to prevent direct contact between the chamber and the glass, a gasket material is compressed between them. The process wets the chamber, the glass, and the gaskets. There is an option to install shields to protect the glass from the process fluid and/or from the ambient environment (e.g. windblown sand). A gauge is referred to as transparent due to the presence of glass panels both in front of and behind the measuring chamber from the perspective of the observer reading the gauge. As a result, light coming from behind illuminates the liquid level indicator. In addition to indicating the level of the liquid, this configuration allows for a visual inspection of the liquid to detect particles and color, for instance. Figure 1.33 shows a transparent level gauge. The liquid level in the gauge glass can be read with the naked eye, which is why it is essential that the material be transparent, allowing the liquid level to be clearly visible. Typically, transparent gauges are used in all standard applications, particularly when the medium is not transparent, and an artificial light source is mounted on the rear of the gauge, which improves the gauge's translucency.

1.8.1.3 Reflex Level Gauges

Reflex-type level gauges operate by detecting the difference in refraction index between fluids and vapors. The refraction and reflection laws of light are used to operate this type of level gauge. In fact, reflex-type glass gauges provide greater visibility of the transition between the liquid level and the gas or vapor above it. Observing a transparent or semi-transparent liquid in a reflex gauge will result

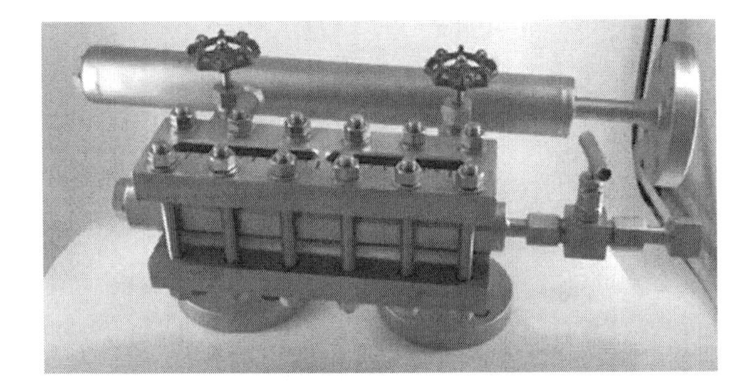

FIGURE 1.33 A transparent level gauge.

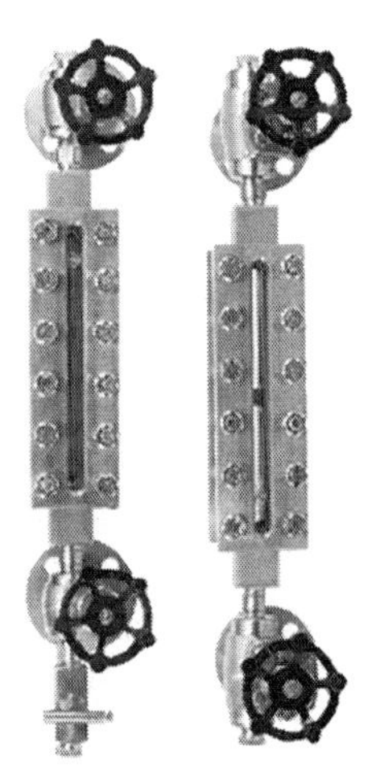

FIGURE 1.34 Reflex level gauges.

in the liquid appearing dark or black since the prism area does not reflect the light back to the viewer. There will be a silvery column above the liquid as light is reflected back from the glass-to-gas interface at the prism area (e.g. there will be a dark column below the indicated level and a silvery column above the indicated level). As a result of the reflection of light back to the viewer, reflex gauges are well suited to being viewed with a flashlight in low-light environments. As a result of the front illumination and prismatic glass, reflex gauges should only be used with clean, clear process fluids and when there is no liquid–liquid interface to observe. They are commonly used in vessels and containers. Normally made of carbon steel or stainless steel, they are durable, can withstand high temperatures and pressures, and are made from a variety of materials. Figure 1.34 shows reflex level gauges.

1.8.1.4 Magnetic Level Gauges

To accurately measure the level of fluid within a vessel, magnetic level gauges use magnets to link the indicator in a gauge to a float inside the vessel. As the float's position changes in the vessel due to fluctuations in the liquid level, the indicator moves up or down by the same amount, ensuring that the liquid level in the measuring chamber is always the same as the fluid level in the vessel itself. Magnetic level gauges operate on the principle of conductivity and consist of three main components: the chamber, the float, and the indicator. There are several types of non-magnetic materials such as stainless steel 316 that can be used to construct the chamber, which is the main component. It is possible to obtain accurate readings from magnetic level gauges by combining proven buoyancy principles with the reliable nature of magnetism.

This magnetic level gauge is designed for the purpose of preventing leakage; ensuring environmental safety; and ensuring safe and trouble-free operation when dealing with chemically aggressive, polluting, hazardous, poisonous, inflammable, or explosive fluids or fluid interfaces with similar optical properties. Magnetic level gauges do not require direct viewing of the level (i.e. no glass is required), so the measuring chamber can be closed, and welded metal construction is generally used. Compared to gauge glass chambers, this significantly broadens the operating temperature range and enhances the ruggedness of the chamber. Because the measuring chamber can have approximately the same thermal expansion coefficient as the vessel, and there is no glass to interface with the metal chamber, a wide temperature range is possible. Figure 1.35 shows magnetic flow meters. The chamber is typically mounted to the side of the vessel, with the liquid level in the

FIGURE 1.35 Magnetic flow meters.

(Courtesy: Shutterstock)

chamber matching the liquid level within the vessel. The magnetic level gauge is similar to a traditional sight glass in this regard, but it may be able to handle some of the more extreme applications, such as those involving highly corrosive or hazardous substances.

1.8.1.5 Bicolor Level Gauges

This design provides a level mainly indicative of steam and other corrosive liquids and gases at high temperatures and pressures. Two flat transparent glasses, together with the gauge body, form the chamber that contains the fluid in these gauges. A special red and green illuminator is mounted on the rear side of the gauge body in order to achieve double illumination. Due to the special light green/red color, the liquid turns green and the steam red for an easy level check. Whenever red radiation hits the water, it is absorbed by one side of the water. When the same light passes through steam, it appears red. In the case of light passing through green filters, the opposite occurs. As a result, users are able to determine the amount of each media that is in the system. Figure 1.36 shows a bicolor level gauge.

1.8.1.6 Radar Level Gauges

Radar level instruments measure the distance between a transmitter (located at a high point) and the surface of a flow located below by measuring time-of-flight. Even under extreme process conditions (pressure, temperature), radar level measurement offers a safe solution. A micropilot works by emitting high-frequency radar pulses that are reflected by the medium surface due to a change in the relative dielectric constant (dc) value. The dielectric constant is defined as the ratio of the electric permeability of the material to the electric permeability of free space. Time-of-flight is directly proportional to distance traveled by the radar pulse. This variable can be used to determine the level of a tank if the geometry of the tank is known. Based on the transmission of radar waves, Figure 1.37 shows a radar-type level gauge indicating that the tank is 50% full.

FIGURE 1.36 A bicolor level gauge.

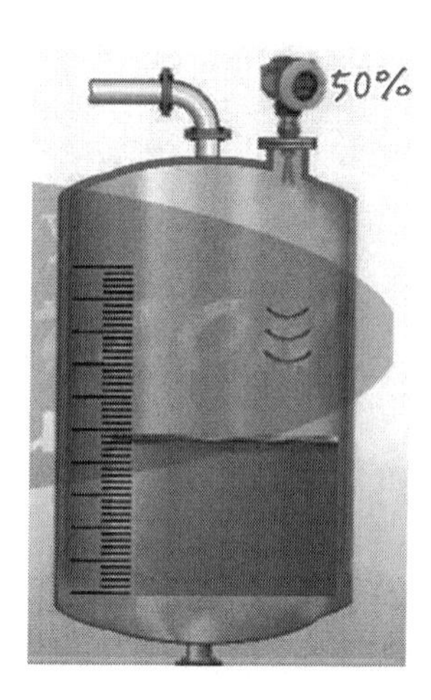

FIGURE 1.37 A radar level gauge.

1.8.1.7 Ultrasonic Level Gauges

Among the major differences between radar and ultrasonic instruments is the type of wave used: sound waves in ultrasonic rather than radio waves. The ultrasonic level gauge is based on the time-of-flight principle. In ultrasonic sensing, a sensor emits ultrasonic pulses, which are reflected by the surface of the media and are detected once again by the sensor. In relation to the distance traveled, the time-of-flight of the reflected ultrasonic signal is directly proportional. It is possible to calculate the level of a tank based on the known tank geometry. The use of ultrasonic level measurement with ultrasonic sensors allows for noncontact, maintenance-free, and continuous level measurement of liquids, pastes, sludges, powders, and coarse bulk materials. The use of ultrasonic devices is particularly advantageous in applications where the measuring device must remain free of contact with the process fluid.

1.8.1.8 Float Level Gauges

One common method of measuring level is through the use of a tape or servo motor connected to a float, as illustrated in Figure 1.38. The height of the float can be read as it moves with the level of the liquid. In float devices, floats are used to indicate the level of liquid in a tank based on their buoyancy. The float is commonly attached to a chain. As the float moves up and down, the chain is attached to a counterweight that indicates the level. Large atmospheric storage tanks often contain these types of devices.

1.8.1.9 Hydrostatic Head Devices

A hydrostatic head device measures the change in static head caused by a liquid. A doubling of liquid depth will result in a doubling of static head (or pressure) in

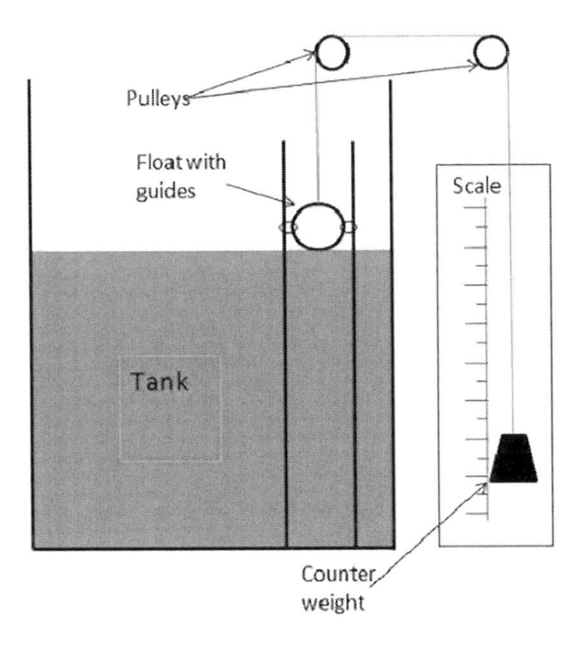

FIGURE 1.38 A float level gauge.

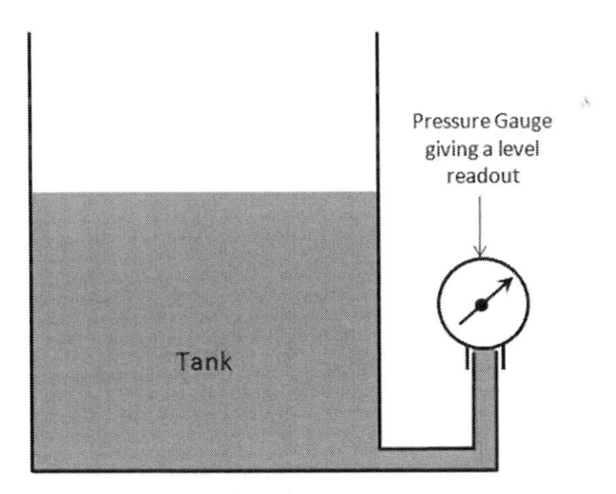

FIGURE 1.39 A hydrostatic head device is used to measure liquid level in the tank.

the tank as well. The bottom of the tank will have a pressure gauge that will register changes in pressure that are proportional to the change in liquid depth. The density of liquid must be accurately measured for accurate level gauges using this method. In Figure 1.39, a hydrostatic head device is shown that is used to measure the liquid level in a tank.

1.9 FLOW MEASUREMENT AND CONTROL

The term "flow" refers to the passage of material from one place to another over a period of time. Flow measurement is the process of measuring fluid flow in the plant or industry. Accurate measurement of flow is essential in a number of process control applications. Flow rates are monitored and controlled at process facilities through the use of measurements. Flow measurements are combined with temperature, pressure, and composition measurements in order to develop material and energy balances for processes. Knowing the amounts of material entering a process and leaving it allows us to calculate material and energy balances quantitatively. It is possible to measure the flow rate of liquids or gases using flow meters.

Flowmeters are devices used to measure the rate and volume of fluid flowing through open or closed conduits. Usually, they consist of a primary and a secondary device. Generally, a primary device is defined as "A device mounted internally or externally to a fluid conduit which produces a signal with a defined relationship to the flow of the fluid in accordance with a known physical law relating the interaction of the fluid with the primary device." According to the definition, a secondary device is "The device that responds to the signal from the primary device and converts it into an output signal that can be translated relative to flow rate and quantity." As required, the secondary device may consist of one or more elements in order to translate the signal from the primary device into standardized or non-standardized display or transmission formats. An example of a flow meter is the orifice flow meter. The liquid or gas whose flow rate is to be determined passes through the orifice plate. Consequently, there is a pressure drop across the orifice plate that varies with the flow rate, which results in a differential pressure between the outlet and inlet segments. In this sense, it consists of an orifice plate, which is the basic component of the instrument. The differential pressure developed across the orifice plate is created when the orifice plate is arranged in a line. There is a linear relationship between this pressure drop and the flow rate of the liquid or gas. Orifice plates are the primary elements of the system, while differential pressure transmitters are the secondary devices that measure the differential pressure caused by orifice plates (see Figure 1.40).

Flow measurement units are selected based on the function and parameters of the measurement. There are a variety of measurement units available depending on the system of measurement and the material being measured. There are a variety of conditions and units that must be used to measure different types of media. Flow can be measured in volume or mass per unit of time. For example, it is common for water resources to be measured in cubic feet per second (cfs), cubic meters per second (cms), or gallons per minute (gpm). It is also possible to express the water flow rate in cubic meters per hour or day. Thus, flow can be expressed mathematically as a ratio of quantity to time. No matter whether we are discussing volumetric or mass flow, the concept remains the same. Simply measuring the

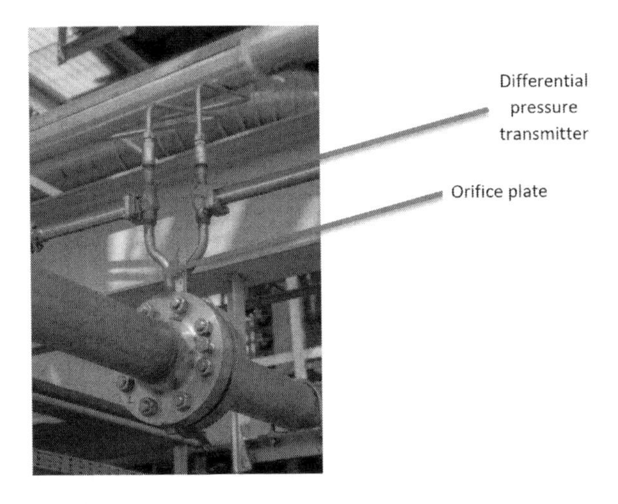

FIGURE 1.40 A flow orifice (primary flow meter) is connected to a differential pressure transmitter (secondary flow meter).

weight (or volume) of stored material over time can provide us with almost any flow rate we desire.

The *volumetric flow rate* is a measure of the amount of fluid passing through a measurement point over a given period of time. A barrel per day is an example of a measurement unit. When the average flow velocity and the inside diameter of the pipe are known, one can calculate the volume flow rate per Equation 1.2.

$$Q = A \times V \tag{1.2}$$

where
Q = volumetric flow rate.
A = cross-sectional area of the pipe.
v = average flow velocity (flow rate).

Mass flow rate refers to the amount of mass passing a particular point over a particular period of time. In a process operation, mass flow rates are used to measure the weight or mass of a substance flowing through it. Using Equation 1.3, the calculation can be performed if the volumetric flow rate and density are known.

$$W = V \times \rho \tag{1.3}$$

where
W = mass flow rate.
Q = volumetric flow rate.
ρ = density.

Example 1.4

Assuming that the vessel's mass decreased from 70,300 kilograms to 70,100 kilograms between 4:05 AM and 4:07 AM, what is the mass flow rate?

$$\text{Answer)}\,\text{Flow rate} = \frac{\textit{mass reduction}\,(kg)}{\textit{time}\,(\text{minute})} = \frac{(70300-70100)\,kilo}{(4:07-4:05)\,(\text{minute})}$$

$$= \frac{200\,kilo}{2\,(\text{minute})} = 100\ \text{kg}\,/\,min$$

There are some flow meters that can measure the volumetric fluid flow rate by measuring the fluid velocity. Speed is measured in terms of distance per second, such as meters per second or feet per second. Due to the fact that the cross-sectional area of a pipe is known, the velocity is then used to calculate the flow rate by multiplying the velocity by the pipe cross section area. In terms of flow rate, m^3/second or ft^3/second represents the amount of fluid volume at a specific location and at a specific time.

1.9.1 Flow Meter Classification

Various methods and technologies can be used to measure flow. In this chapter, eight categories of flow meters are discussed: mechanical flowmeters, pressure drop-based flowmeters, vortex flowmeters, optical flowmeters, thermal flowmeters, ultrasonic flowmeters, electromagnetic flowmeters, and mass flowmeters.

1.9.1.1 Mechanical Flow Meters

A mechanical flow meter consists of a moving part or rotating device such as a propeller or paddle wheel. By passing liquid through the mechanical flowmeter, the moving part is induced to rotate. A flowmeter generates a flow rate based on the movement of its moving parts. The principle of mechanical flowmeters can be applied to a variety of instruments. Among these are piston meters, turbine flow meters, variable area meters, woltman meters, single jet meters, paddle wheel meters, rotating disc meters, oval gear meters, and pelton meters.

1.9.1.1.1 Piston Flow Meters

A piston flow meter consists of a rotating piston inside a chamber. The volume of the container is known. The piston displaces the liquid's volume as it passes through the chamber. During the process of filling and emptying the chamber, the piston rotates. A total volume of liquid is determined by counting the number of rotations. The piston meter operates on the principle of positive displacement flowmeters. Low-volume fluids can be measured using this device. Figure 1.41 shows a piston flow meter.

1.9.1.1.2 Turbine Flow Meters

The turbine flow meter was invented by Reinhard Woltman in the 18th century and is suitable for use with both liquids and gases. This device consists of a

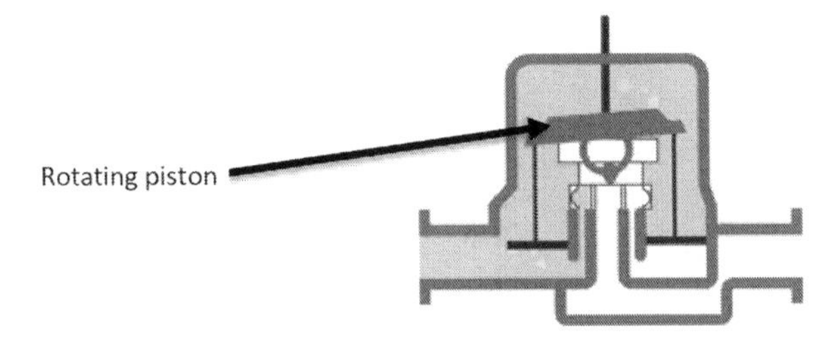

FIGURE 1.41 A piston flow meter.

multi-bladed rotor, also called turbine blades, mounted at right angles to the fluid flow and suspended within it. The diameter of the rotor is slightly less than the inside diameter of the metering chamber, and its rotation speed is proportional to the volumetric flow rate. The flowing fluid engages the rotor, causing it to rotate at an angular velocity proportional to the rate of fluid flow. A turbine flow meter uses the mechanical energy of the liquid to rotate a rotor within the flow stream. As the fluid travels through the meter, there is a direct relationship between rotational speed and fluid velocity. It is important to note that turbine meters are an effective solution in applications involving constant, high-speed flows in which accuracy is paramount. As shown in Figure 1.42, there are a variety of turbine flow meters.

1.9.1.1.3 Variable Area Flow Meters

Variable area flow meters are very simple yet versatile flow measurement devices that can be used with all types of liquids, gases, and steam. Variable area meters are generally configured as cone-and-float meters, also known as rotameters. In general, a variable area meter is composed of a tapered tube (usually glass) that contains a self-centering float that is raised by the flow and lowered by gravity. A float is a solid object suspended in a tube that serves as a flow indicator. Figure 1.43 shows a variable area flow meter.

1.9.1.1.4 Woltman Flow Meters

The Woltman meter (see Figure 1.44) is a turbine flowmeter consisting of a rotor with helical blades that are inserted axially into the meter. Through the use of a turbine, it measures the velocity of the liquid flowing. Using roller counters in cubic meters, the volume is calculated mechanically based on the known section area velocity.

1.9.1.1.5 A Single-Jet Flow Meters

A single-jet meter creates a water jet from a single port, as its name suggests. Upon starting the turbine, it transmits the motion to the display mechanism, which allows us to measure the volume of water passing through the meter. In order to measure the volume of water flowing through the meter, the turbine starts rotating and transmits motion to the display mechanism. Water meters with a single jet are simple and include an internal strainer to prevent the jet from becoming clogged. A multi-jet meter differs from a single-jet meter in that it has multiple ports that surround an internal chamber to create a flow of water against the turbine.

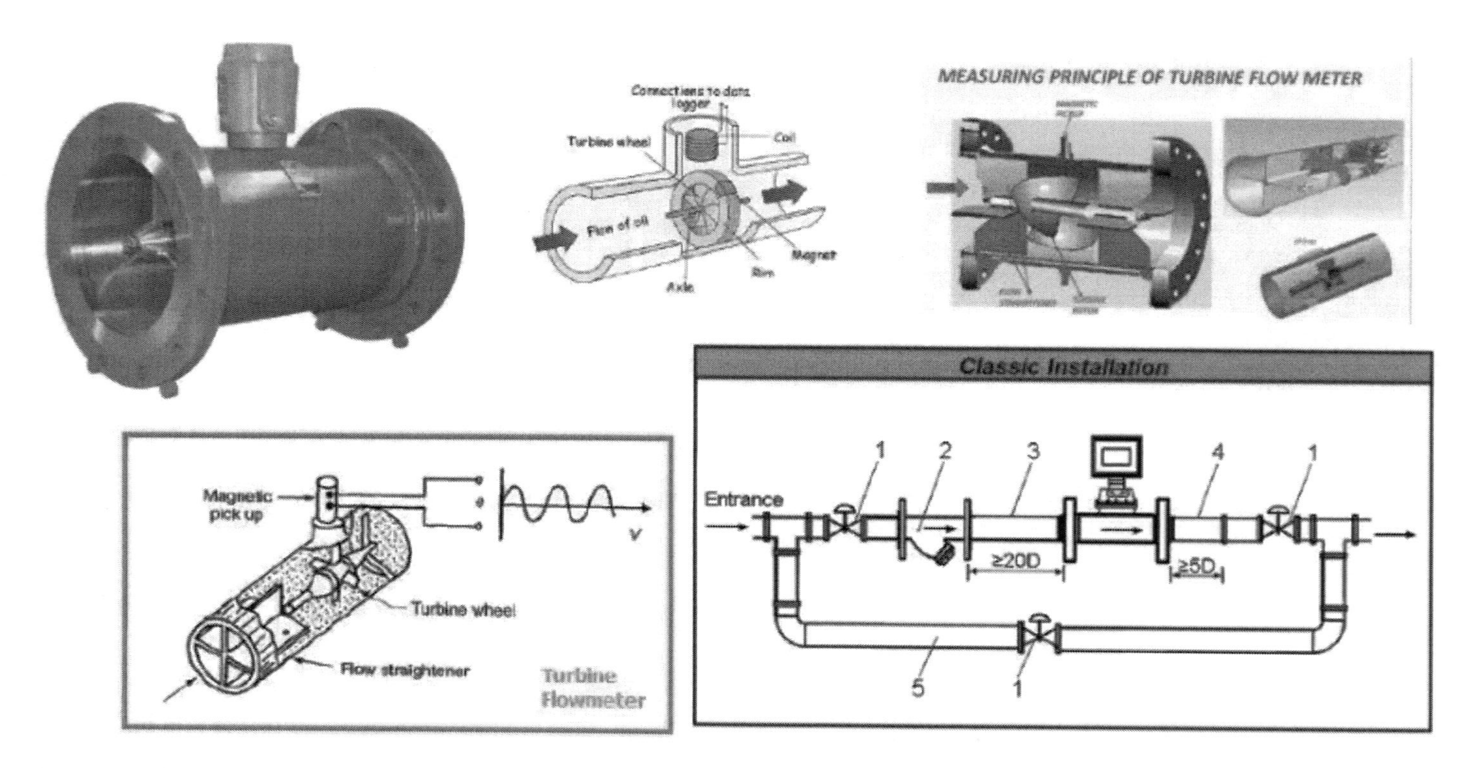

FIGURE 1.42 Various turbine flow meters.

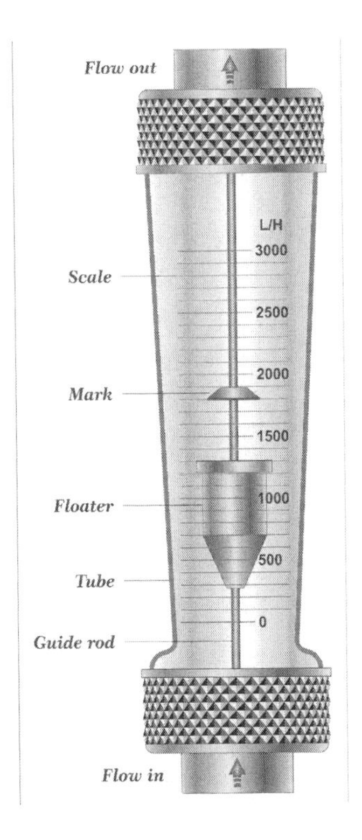

FIGURE 1.43 A variable area flow meter.

(Courtesy: Shutterstock)

FIGURE 1.44 A Woltman flow meter.

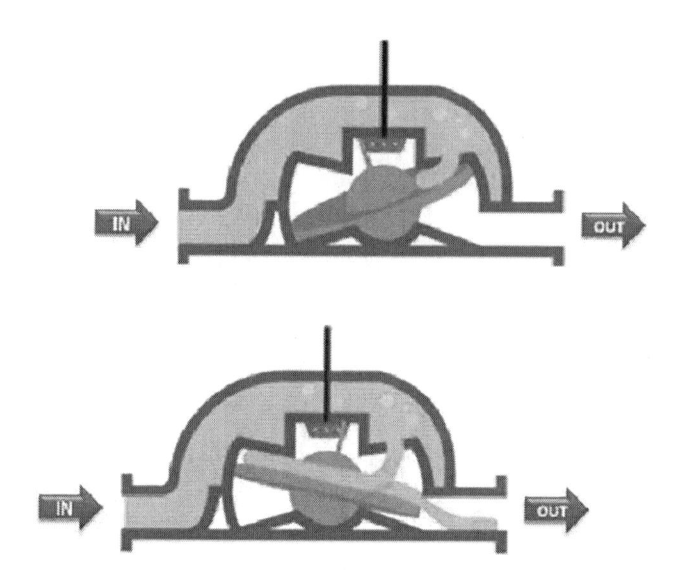

FIGURE 1.45 A nutating disk flow meter.

1.9.1.1.6 Paddle Wheel Flow Meters

A paddle wheel meter is composed of a paddle that rotates as the fluid passes through the pipe. There is a proportional relationship between the speed of rotation of the paddle and the flow of liquid inside the pipe.

1.9.1.1.7 Nutating Disk Flow Meters

A nutating disc flow meter is one of the most common types of positive displacement flow meters. A disc is mounted on a ball in the center of the device, as illustrated in Figure 1.45. Fluid entering the chamber causes the disc to wobble (nutate), causing the displaced volume to be transferred to the register.

1.9.1.1.8 Oval Gear Flow Meters

Two oval-shaped precision wheels are used in an oval gear meter, as shown in Figure 1.46. The wheels of the meter rotate proportionally to the amount of fluid passing through them. In order to determine the volume, the number of rotations is used.

1.9.1.1.9 Pelton Flow Meters

One type of turbine flow meter is one in which the rotor rotates in the same direction as the flow rather than perpendicular to it. A paddle wheel flow meter and a pelton wheel flow meter are examples of this type of turbine meter. There are two types of flow meters that are particularly suitable for liquids. As a result of a rotating Pelton wheel, the Pelton meter converts its mechanical action into a form that can be read by the user (a flow rate that can be read).

FIGURE 1.46 An oval gear flow meter.

1.9.1.2 Pressure Drop–Based Flow Meters (Differential Pressure Flow Meters)

Differential pressure flow meters measure the volume flow of gases, liquids, and steam. In particular, they are used in situations where high pressure, high temperature, or a large diameter are relevant. Their main application areas are in the chemical, oil, gas, and power industries. The differential pressure flow meter is a measurement technology that uses Bernoulli's principle to measure the flow of liquid, steam, or gas in pipes. Bernoulli's principle is a fundamental concept in fluid dynamics that relates pressure, speed, and height. Bernoulli's principle states that an increase in speed of a fluid is accompanied by a decrease in static pressure. A differential pressure meter based on Bernoulli's principle provides more accurate, reliable readings than many alternatives since the principle describes the relationship between the velocity of a fluid and its pressure. With the introduction of a constriction or obstruction in the pipe, differential pressure (DP) flow meters can precisely measure the flow of fluid within the pipe. With an increase in pipe flow, a greater pressure drop is generated. The result is a nearly immediate response time to any changes in pressure, regardless of the flow velocity or other characteristics of the flow. There are three main types of differential pressure flow meters: orifice plates, pitot tubes, and Venturi meters. As shown in Figure 1.47, differential pressure flow meters come in a variety of designs.

1.9.1.2.1 Orifice Plates

Orifice plates consist of metal disks with concentric holes, which are inserted into the pipe carrying the flowing fluid. A wide range of sizes and shapes are available for orifice plates, which are simple, inexpensive, and easy to use. Orifice plates are typically sandwiched between two flanges of a pipe joint, making them easy to install and remove. In the vena contracta, the cross-sectional area of the fluid flow profile is reduced to a minimum after passing through the orifice, and this area has the lowest fluid pressure (see Figure 1.48).

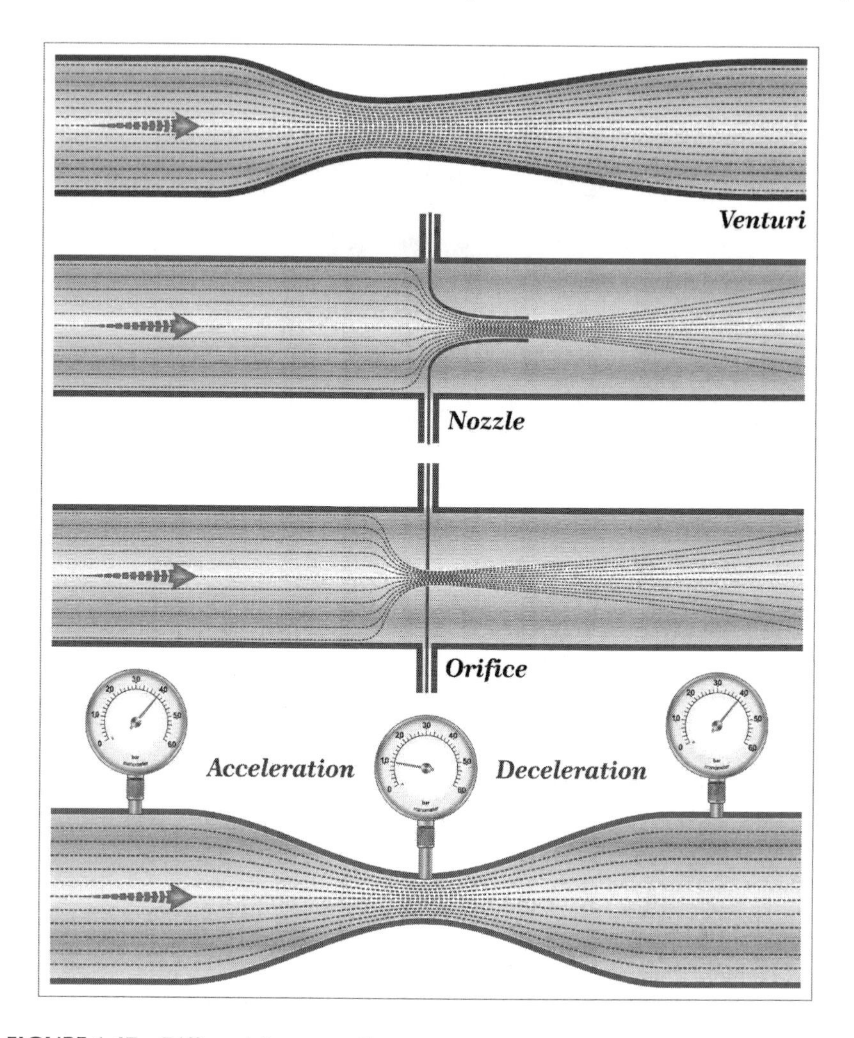

FIGURE 1.47 Differential pressure flow meters.

(Courtesy: Shutterstock)

1.9.1.2.2 Pitot Tubes

Henri Pitot invented pitot tubes in 1732 for measuring the flow velocity of fluids. A pitot tube is a differential pressure flow meter that measures flow velocity using no moving parts. A pitot tube is a common type of insertion flowmeter. A more recent design incorporates a straight or L-shaped, multi-chamber tube with multiple holes in the front and rear chambers, perpendicular to the flow direction. Each chamber emerges separately from the pitot tube at the top for connection to the pressure instrument. In the case of pitot tubes that are installed in conduits with known cross-sectional areas, it is possible to calculate the volumetric flow by

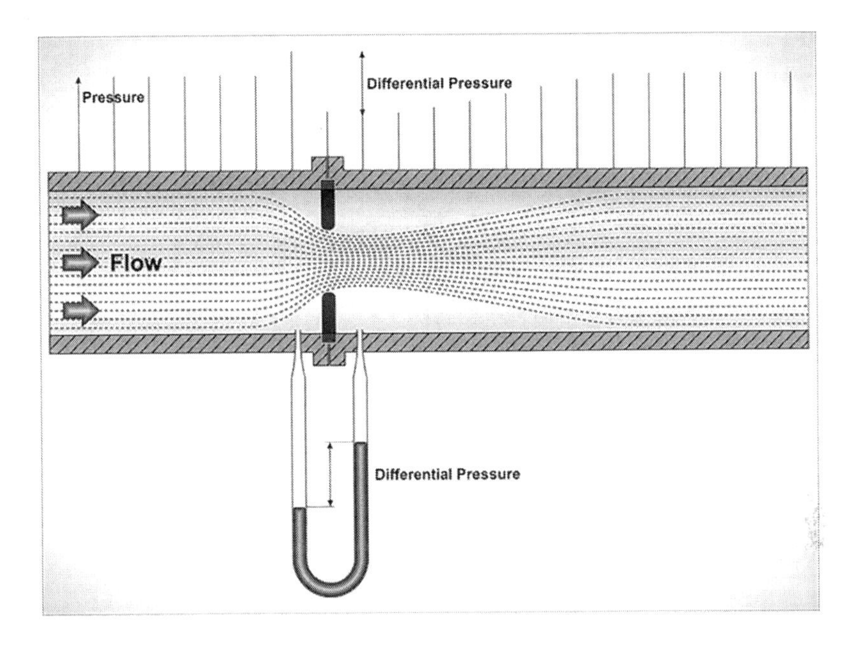

FIGURE 1.48 An orifice plate installed in piping.

(Courtesy: Shutterstock)

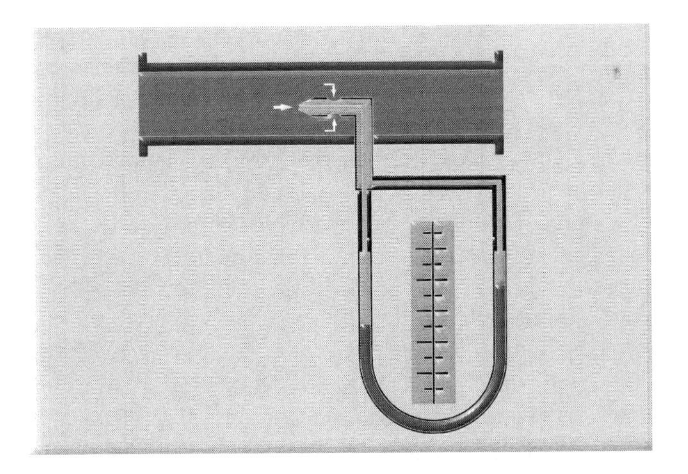

FIGURE 1.49 A pitot tube flow meter connected to a differential pressure gauge.

multiplying the measured flow velocity by the cross-sectional area of the conduit. As long as the density of the fluid is known, a differential pressure transmitter can measure the DP signal from a pitot tube and convert it to a flow velocity. A pitot tube flow meter is created by combining a pitot tube and a differential pressure transmitter (see Figure 1.49).

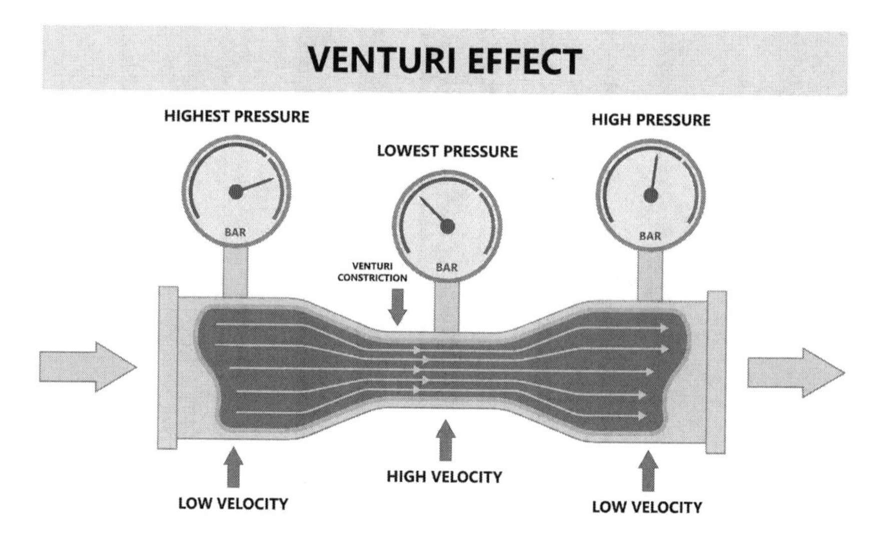

FIGURE 1.50 A Venturi flow meter.

(Courtesy: Shutterstock)

1.9.1.2.3 Venturi Flow Meters

A venturi flow meter measures the difference in pressure between two different locations in a pipe in order to generate a flow measurement (see Figure 1.50). Pressure differences are created by constricting the diameter of the pipe, which results in an increase in flow velocity and a corresponding drop in pressure. Through these changes in fluid velocity and pressure, the flow rate can be determined. There is a pressure drop between the entrance and throat (middle section) of a Venturi flow meter when it is installed in a pipe carrying the fluid whose flow rate is to be measured. By calibrating the differential pressure sensor, it is possible to measure the flow rate based on the pressure drop measured by the sensor. All differential pressure flow meters operate based on the Bernoulli equation, which states that as the speed of flow of a fluid increases, pressure loss occurs. A classical Venturi flow meter is characterized by a long flow element with a tapered inlet and a diverging outlet. At the entrance, the inlet pressure is measured, and at the throat section, the static pressure is measured. In this element, the pressure taps are connected to a common annular chamber so that the pressure reading is averaged over the entire circumference. Classical Venturis are limited to the use of clean, noncorrosive liquids and gases.

1.9.1.3 Vortex Flow Meters

Vortex flow meters have the advantage of measuring the flow of all three phases of a fluid, namely gas, liquid, and steam. The fact that they do not have moving parts makes them cost effective in terms of maintenance and in measuring fluids that can generate problems due to moving parts. Vortex flowmeters operate on the principle

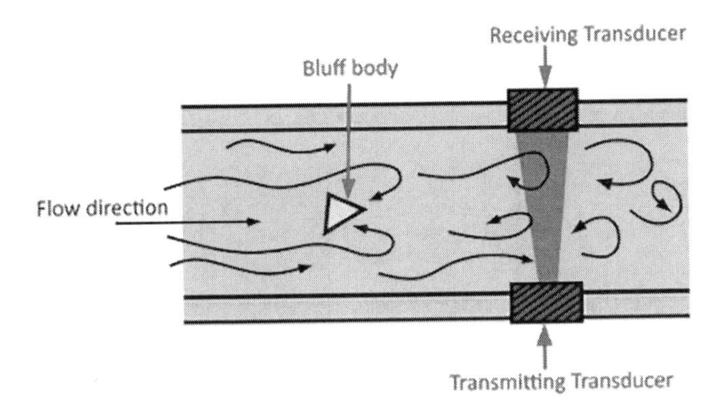

FIGURE 1.51 A vortex flow meter.

of vortex shedding. As a fluid passes an obstacle, oscillating vortices are observed downstream of the obstacle, as illustrated in Figure 1.51. Pulse sensors measure the frequency of these oscillations, which is proportional to the volume flow.

1.9.1.4 Optical Flow Meters

In optical flow meters, light is used to measure the flow rate, which is based on the principle of optics. A relatively recent development in industrial flow measurement is the optical flow meter, which uses light to measure the velocity of fluids through pipes. The flow rate of a stream can be determined by measuring the time of flight of light.

1.9.1.5 Thermal Mass Flow Meters

A thermal mass flow meter is all about the transfer of heat that is involved in thermal mass flow measurement. The heat will be absorbed into the fluid as a result of a flow of gas or liquid passing over a heated surface. As a result of such a flow, the heated surface is cooled. As the flow increases, the heated surface will be cooled to a greater extent. Thus, it is possible to measure the flow rate by measuring the temperature of the heated surface.

1.9.1.6 Ultrasonic Flow Meters

Ultrasonic flow meters measure the velocity of a fluid using ultrasound to calculate volume flow. By averaging the difference in measurement transit time between pulses of ultrasound propagating into and against the direction of the flow, the flow meter can measure the average velocity along the path of an emitted beam of ultrasound. Flow meters that use ultrasonic technology measure the velocity of fluids without contacting them. They are clamp-on devices that attach to the exterior of the pipe (and fit a variety of pipe sizes) and enable measurement of corrosive liquids without damaging the ultrasonic sensor. Two opposing probes are placed upstream and downstream to achieve the ultrasonic transit time principle. An ultrasonic wave is transmitted and received alternately by each probe of a

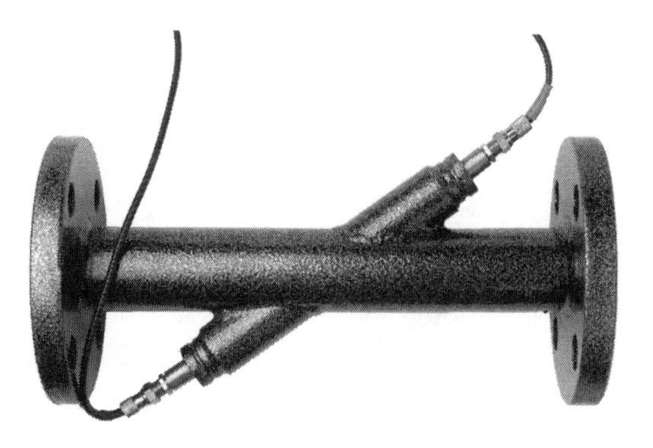

FIGURE 1.52 An ultrasonic flow meter.

(Courtesy: Shutterstock)

chord. Waves travel downstream–upstream and upstream–downstream at different speeds (when the fluid is moving, the downstream–upstream time is greater than the upstream–downstream time). By determining the average flow velocity along the chord and the difference between the two times, the volume flow rate can be calculated. It is shown in Figure 1.52 that a pipe section with ultrasonic sensors is used for measuring liquid flow.

1.9.1.7 Electromagnetic Flow Meters

In accordance with Faraday's law of electromagnetic induction, electromagnetic flow meters are a type of velocity or volumetric flow meter that induces a voltage when a conductor moves through a magnetic field. As a conductive liquid moves through a magnetic field, the voltage induced by the liquid is directly proportional to its speed. A diametrically opposed pair of electrodes is used to measure the induced voltage. Using the voltage, it is possible to determine the velocity of the fluid and its volumetric flow rate.

1.9.1.8 Coriolis Flow Meters

Flowmeters that use Coriolis technology consist of a tube (or tubes) that is subjected to vibration. A fluid (gas or liquid) passing through this tube causes the vibrating tube to distort due to the inertia of the mass flow, resulting in a phase shift between the inlet and outlet. Measurement of the phase shift is possible and is proportional to the mass flow.

1.9.2 Flow Measurement Device Selection Criteria

It can be a very challenging task to select a flowmeter due to the wide range of technologies and application requirements that must be met. There are several approaches to selecting flowmeters that begin with an elimination process. This process reduces the number of available flowmeters to a few that are suitable for the application. However, this does not imply that the "acceptable" choices are necessarily the "perfect" ones. In selecting a flowmeter, the following factors are considered: application fundamentals, specifications, safety considerations, metallurgy, installation considerations, maintenance and calibration, compatibility with existing process instruments, custody transfer concerns, economic considerations, and technical direction.

In terms of application fundamentals, selecting a flowmeter begins with a very simple yet understated question—what does the instrument need to accomplish? Nevertheless, using that simple question as a starting point, one can use approaches to flowmeter selection that are heavily influenced by an understanding of its intended use. Specifications for flow meters refer to the factors that determine the selection and design of flow meters, such as the type of service, accuracy, pressure, temperature, and pipe size. Considerations regarding safety that are important when selecting a flowmeter are described as follows: providing protection to the flowmeter, hazardous area requirements, and personnel protection. For example, strainers may be installed prior to flow meters to protect them from debris. It is imperative that the necessary precautions be taken when installing or using a flowmeter in a hazardous location or when measuring an explosive fluid. As a result of good design practices, proper installation practices, and proper maintenance procedures, personnel are protected. Installation considerations for flow meters include upstream and downstream piping requirements. The length of the upstream piping is often determined by the type of meter. To prevent erosion inside the meter, it is also necessary that the upstream pipe be straight in order to straighten the flow and provide more accurate measurement. The downstream piping requirements are less obvious. In downstream piping, control valves or obstructions may cause opposite effects to the flow direction. When the effects are large enough, they can back up to the meter, enter the meter, and affect the measurement.

QUESTIONS AND ANSWERS

1. Which statement is correct regarding the measurement system?
   A. The primary transmitter is the starting point of a measurement system.
   B. Through the use of signal processing elements, it is possible to improve the quality of a measurement system's output.
   C. The primary sensor and variable conversion element are sometimes combined to form a transmitter.
   D. It is possible for signal processing to be integrated into a transmitter, which is then referred to as a transducer.

   Answer) Option A is incorrect since the measurement system is initiated with sensors. The correct answer is option B. Option C is incorrect, since the combination of a sensor with a variable conversion element should be referred to as a transducer and not a transmitter. It is incorrect to choose option D because, when signal processing is integrated into a transducer, the device is referred to as a transmitter.

2. In comparison to deflection instruments, what is the main advantage of null instruments?
   A. Faster response
   B. Higher accuracy
   C. Lower sensitivity
   D. All three choices are correct

   Answer) Option B is the correct answer.

3. In the case of a pressure gauge with an accuracy of plus or minus 5% and a range between 0 and 1 bar, what is the maximum error?
   A. 0.02 bar
   B. 0.03 bar
   C. 0.04 bar
   D. 0.05 bar

   Answer) Maximum error = 5% × 1 = 0.05 bar
   Therefore, option D is the correct answer.

4. What type of level gauge converts the pressure of a liquid level column into a level reading?
   A. Float level gauge
   B. Hydrostatic head level gauge
   C. Bi-color level gauge
   D. Ultrasonic level gauge

   Answer) Option B is the correct answer.

5. When comparing thermocouples with RTDs, which statement is correct?
   A. Thermocouples are more expensive than RTDs.
   B. Thermocouples are more accurate than RTDs.
   C. Thermocouples have faster response to temperature change than RTDs.
   D. All three choices are incorrect.

   Answer) Option C is the correct answer.

6. Identify the correct statement regarding flow meters.
   A. There are three types of mechanical flow meters: orifice plate, pitot tube, and Venturi meter.
   B. It is always necessary to express flow measurements in terms of mass flow rate.
   C. By measuring the flow rate of the liquid, the turbine flow meter determines its volume. When fluid passes through the turbine blades, the turbine blades begin to rotate. It consists of freely suspended turbine blades.
   D. A single-jet meter is a type of rotameter that consists of a tube and float. A float is a solid object that is suspended within a tube and

serves as a flow indicator. As the liquid flows upward through the tube, the differential pressure created across the float is measured.

Answer) Option A is incorrect since orifice plates, pitot tubes, and Venturi meters are considered pressure differential flow meters rather than mechanical flow meters. Option B is incorrect because mass and volume can both be measured for the flow. The correct answer is option C. Option D is incorrect because the given definition refers to a variable area flow meter and not to a single-jet meter.

7. What is not an advantage of noncontact temperature sensors?
   A. The process is fast (within a few milliseconds). As a result, a significant amount of time can be saved, allowing for more measurements and data accumulation.
   B. It is possible to measure the temperature of objects that are hazardous or physically inaccessible.
   C. Measurements of high temperatures (up to 3000°C) are not problematic.
   D. It is used to protect temperature sensors, such as thermocouples, thermistors, and bimetal thermometers.

   Answer) Option D describes the application of thermowells and does not apply to noncontact temperature sensors.

8. Which of the following flow meters is classified as a differential pressure transmitter?
   A. Pitot tube
   B. Vortex flow meter
   C. Turbine flow meter
   D. Paddle wheel meter

   Answer) Option A is the correct answer.

9. Which flow meter operates according to Faraday's law of electromagnetic induction?
   A. Coriolis flow meter
   B. Electromagnetic flow meter
   C. Orifice flow meter
   D. Turbine flow meter

   Answer) Option B is the correct answer.

10. In selecting flow meters, what is the most important parameter?
    A. Safety
    B. Size
    C. Pressure and temperature
    D. All answers are correct

    Answer) Option D is the correct answer.

FURTHER READING

1. Liptak, B. G. (2003). *Instrument engineers' handbook—process measurement and analysis*. Vol. 1. 4th ed. Boca Raton: CRC Press.
2. Control Automation. (2023). *Force-balance pressure transmitters*. [online] available at: https://control.com/textbook/continuous-pressure-measurement/force-balance-pressure-transmitters/ [access date: 8th July, 2023]
3. Electrical Deck. (2021). *Classification of measuring instruments—indicating, integrating & recording*. [online] available at: www.electricaldeck.com/2021/04/classification-of-measuring-instruments.html [access date: 3rd July, 2023]
4. Fisher & Emerson Automation Solutions. (2017). *Control valve handbook*. 5th ed. New York, NY: Fisher & Emerson Automation Solutions.
5. Grodzinsky, E., & Sund Levander, M. (2020). History of the thermometer. In *Understanding fever and body temperature: A cross-disciplinary approach to clinical practice*, pp. 23–35. London: Palgrave Macmillan.
6. Instrument Tools. (2023). *Types of level gauges*. [online] available at: https://instrumentationtools.com/types-level-gauges/ [access date: 6th July, 2023]
7. Marlin, T. (2000). *Process control: Designing processes and control systems for dynamic performance*. Boston: McGraw Hill. ISBN: 0-07-039362-1
8. McIntyre, C. (2011). *Using smart instrumentation*. [online] available at: www.controleng.com/articles/using-smart-instrumentation/ [access date: 3rd July, 2023]
9. Morris, A. S., & Langari, R. (2012). *Measurement and instrumentation: Theory and application*. Waltham, MA: Academic Press. ISBN: 0128008849
10. Mulindi, J. (2023). *Null vs deflection types instrument*. [online] available at: www.electricalandcontrol.com/null-vs-deflection-type-instruments/ [access date: 3rd July, 2023]
11. Nakra, B. C., & Chaudhry, K. K. (2004). *Instrumentation, measurement and analysis*. 2nd ed. New Delhi: Tata McGraw-Hill.
12. Palazoglu, A., & Romagnoli, J. (2006). *Introduction to process control*. Boca Raton: Taylor & Francis. ISBN: 0-8493-3696-9
13. Porter, M. A., & Martens, D. H. (2002, January). Thermowell vibration investigation and analysis. In *ASME pressure vessels and piping conference*. Vol. 46571, pp. 171–176. New York, NY: ASME.
14. Rajita, G., Banerjee, D., Mandal, N., & Bera, S. C. (2015). Design and analysis of Hall effect probe-based pressure transmitter using bellows as sensor. *IEEE Transactions on Instrumentation and Measurement*, 64(9), 2548–2556.
15. Roux, S. (2022). *What is flow measurement*. [online] available at: https://faure-herman.com/articles/what-is-flow-measurement [access date: 11th June, 2023]
16. Saito, K., Sawahata, H., Homma, F., Kondo, M., & Mizushima, T. (2004). Instrumentation and control system design. *Nuclear Engineering and Design*, 233(1–3), 125–133.
17. Sotoodeh, K. (2020). Challenges associated with the bypass valve of control valve in a sea water service. *Journal of Marine Science and Application*, 19(1), Springer. https://doi.org/10.1007/s11804-020-00132-8

2 Piping Components

2.1 INTRODUCTION

Pipelines and piping are both used to transport liquids, gases, and slurries. The piping system is typically connected to a variety of equipment and carries fluids that will be processed within that equipment inside a complex network. When pipelines are used to deliver fluids or end products to other facilities for further processing, they are typically straight and supply the feed for further processing. There are many straight pieces of pipe that are welded together over a long distance to form a pipeline. Figure 2.1 illustrates China's West–East gas pipeline, which is approximately 9 kilometers in length. In contrast, piping is a system of pipes and fittings within the boundaries of a refinery or petrochemical plant, as shown in Figure 2.2. While pipelines may be above or below ground, piping is almost always above ground.

In this chapter, we discuss bulk piping components such as pipe, pipe fittings, flanges, gaskets, and bolts, as well as special piping components such as strainers and steam traps. The purpose of pipe fittings, also known as pipe connectors, is to join one pipe to another in order to extend the run or change the flow direction or size of a piping system. The purpose of flanges is to connect pipes, valves, fittings, and special items to each other. Bolting is used to connect flanges, and gaskets or other sealing devices are often used to seal them together.

2.2 PIPING PARAMETERS

In terms of pipes, there are a number of parameters that are important. These include the pipe size, the thickness of the pipe, its length, the ending, the standard, the material, and the manufacturing method.

2.2.1 Piping Size

There are several attributes and parameters that must be specified when defining a pipe in piping material specification. The first and most important parameter is the pipe's size. Typical pipe sizes are ½", ¾", 1", 1 ½", 2", 3", 4", 6", 8", 10", 12", 14", 16", 18", 20", 24", and more. Pipe sizes such as 1 ¼", 2 ½", 3 ½", 5", and 7" are not used. The pipe is designated by its nominal size, which corresponds to its outside diameter. For sizes up to and including size 12", the outside diameter of the pipe is greater than the nominal pipe size (NPS). In sizes larger than 12", the NPS and outside diameter of the pipe are equal. The nominal pipe size is a North American standard for identifying pipes of different sizes. According to technical terms, the NPS is a non-dimensional measurement that refers to the pipe's

DOI: 10.1201/9781003465881-2

FIGURE 2.1 Part of the West-East gas pipeline in China.

(Courtesy: Shutterstock)

FIGURE 2.2 Piping system in a refinery.

(Courtesy: Shutterstock)

diameter only roughly. Pipes outside North America are identified by diameter nominal (DN), a dimensionless value that roughly corresponds to the pipe's outside diameter in millimeters. There is the possibility that you will hear the terms "large bore" and "small bore" when working on a project. Typically, pipes with sizes between 1 12" and 1 12" are considered small bores, while pipes with sizes of 2" and above are considered large bores.

2.2.2 Piping Thickness (Schedule)

It is important to note that while the nominal pipe size identifies the outside diameter of the pipe, more information is needed in order to identify specific pipes. Pipe schedules refer to the thickness of the pipe wall, which directly impacts the internal dimensions and weight of the pipe. In order to determine how much internal pressure the pipe can withstand, it is important to know the thickness of the wall. A piping engineer's most important and fundamental task is to calculate the wall thickness of pipelines and piping. Pipelines and piping in the oil and gas industry handle high-pressure fluids at specific temperatures. It should be noted that different ASME B31 piping and pipeline codes provide different methods and equations for calculating pipe and pipeline wall thickness. Schedule = 1000 × (P/S), where P is the internal service pressure of the pipe (pound per square inch), and S is the ultimate tensile strength of the pipe material (psi). Common schedule numbers in the pipe standards are 5, 10, 20, 30, 40, 60, 80, 100, 120, 140, and 160. Additionally to the schedule numbers, you may also see Standard (STD), Extra Strong (XS), and Double Extra Strong (XXS) in piping standards such as American Society of Mechanical Engineers (ASME) B36.10 and B36.19. Prior to the implementation of the schedule system in 1927, these terms were used. It is customary to add an "S" to the end of the schedule number of stainless-steel pipes, for example, Schedule 40S. The thickness of the pipe affects the piping inside diameter. In fact, the outside diameter of the pipe is constant, and its inside diameter (ID) is calculated by extracting the double thickness from the outside diameter (OD) of the pipe $(OD - 2 \times Thickness = ID)$. Consequently, two pipes with the same external diameter but different thicknesses have different internal diameters in such a way that the thicker pipe has a smaller internal diameter.

2.2.3 Pipe Length

A pipe's length, which could be a single random length (approximately 6 m on average) or a double random length (12 m on average), should be given to the pipe manufacturer in the engineering documents.

2.2.4 Pipe Ending

While size is an important consideration when choosing piping and piping components, pipe ends are equally important to ensure a proper fit, a tight seal, and optimal performance. Depending on the end of the pipe, it may have a plain end (PE), a threaded end (TE), or a bevel end (BE). Pipe ends with plain edges are characterized by a sharp, abrupt edge at the end of the pipe run (see Figure 2.3). It is necessary for a pipe to have a cut of 90° running perpendicular to the edge of the pipe for it to be classified as plain ended. Due to the fact that this end cannot be connected directly, other fittings should be used in order to connect two pipes with plain ends. A plain end pipe is a male connection connected to a fitting with a female connection by socket welding. In the case of corrosive media such as seawater, socket welding is not an appropriate method of joining, due to

FIGURE 2.3 A plain-end pipe.

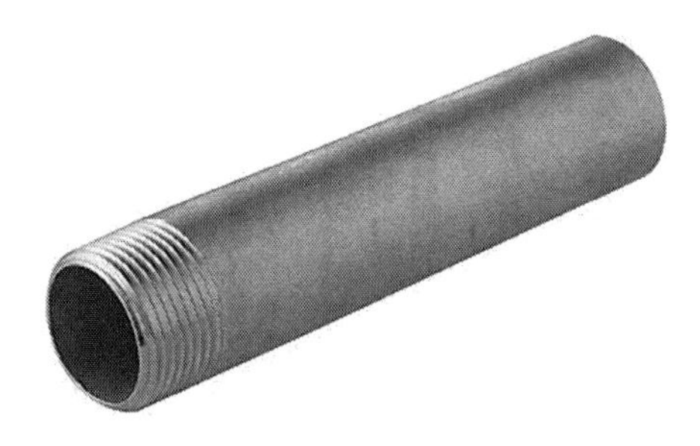

FIGURE 2.4 A threaded-end pipe.

the possibility that the fluid could get trapped within the gap in the socket weld and cause corrosion of the crevices. The term crevice corrosion refers to a local attack on a surface where there is a gap between two surfaces. A socket weld is used to attach piping with plain ends by inserting the pipe into a recessed area of the fitting. Following the insertion of the pipe into the fitting's female area, called the socket end, the outer area of the pipe is fillet welded to the fitting. In the fitting, there is a gap between the pipe and the socket female end. There is a 1.6-mm gap in the piping due to the possibility of thermal expansion during the fillet welding process. The gap is considered a weak point since corrosive fluid could accumulate there and cause crevice corrosion. Due to this, socket welding is not recommended for services that could cause crevices and pitting corrosion, such as seawater. Pipes with plain ends are typically used for small diameters of 2" or less.

Threaded piping (see Figure 2.4) has male threads, which are inserted into fittings with female threads. The use of threaded joints is typically restricted to small-diameter pipes up to 2" in diameter. A threaded joint is not considered

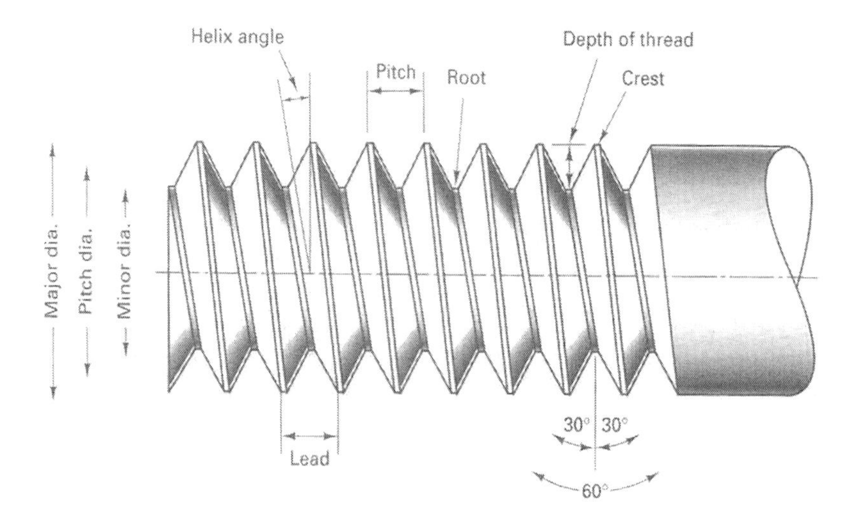

FIGURE 2.5 Pipe thread expressions and geometry.

robust and is not typically used in hazardous applications such as those involving flammables, high pressure, and corrosives.

Figure 2.5 illustrates different expressions and geometry for pipe thread connections, as well as "depth of thread." As can be seen in the picture, "*depth of thread*" decreases piping thickness and therefore should be added to the thickness of threaded-end pipes. Thread *pitch* refers to the distance between adjacent threads on a screw. It can also be referred to as TPI, or threads per inch. To determine thread pitch, you can use a pitch gauge, which gives you an approximate measurement based on established thread standards. The *lead* is the linear distance that is traveled by a nut per revolution of a screw. As the name implies, a thread's *crest* is the most prominent part of the thread, whether it be internal or external. *Roots* are the bottom of the groove between the two flanking surfaces of the thread, whether internal or external.

As the most robust piping connection, buttwelds are bevel ended in accordance with ASME B16.25 standards. Weld end bevel dimensions are provided in this standard. It is possible to prepare the bevel ends by machining. The most commonly used ends on the market are beveled ends. Besides its role in fusing pipe ends, beveling is also a popular alternative for safety and design reasons. By beveling the ends instead of having a 90° plain end, the pipes are better able to fuse together. Typically, beveled ends are used in buttwelding, where pipes are cut at specific angles at the ends. As a result of this shaping, fittings and flanges can be added over the welded area, thereby increasing the strength of the pipe. Figure 2.6 illustrates how to bevel the ends of pipes and fittings in accordance with ASME B16.25.

Two scenarios of bevel-end preparation based on piping thickness are shown in Figure 2.6, one for wall thickness (t) up to and including 22 mm, and one for

FIGURE 2.6 Seamless pipes.

(Courtesy: Shutterstock)

thicknesses greater than 22 mm. On average, a 1.6-mm root face is prepared, and the bevel end fitting or pipe has 37.5° angle on average to the vertical line for wall thicknesses up to and including 22 mm, according to ASME B16.25. The angle should be reduced to 10° on average for thickness values above 22 mm on the extra thickness over 19 mm on average. A narrow gap welding procedure is an advanced welding end preparation method in which the angle of the bevel end fitting is 7° with respect to the vertical line, as illustrated in this example. Among the advantages of narrow gap welding are the use of fewer welding electrodes, the speed of the welding process, and the use of less heat. The narrow gap welding process is based on industrial practices rather than the ASME B16.25 standard.

2.2.5 Pipe Standards

It is possible for the pipe to be manufactured in accordance with ASME B36.10, 19, 10M, or 19M standards. For carbon steel piping, the ASME B36.10 or 10M standard is commonly used. For stainless steel piping, the ASME B36.19 or 19M standard is typically used.

2.2.6 Pipe Material

A piping material grade, which is normally determined by the American Society for Testing and Materials (ASTM), is the other designation for piping materials. For example, A 106 Gr. B represents seamless, carbon steel pipe. An ASTM grade indicates not only the material of the pipe but also the method by which the pipe was manufactured.

For piping material and standard designations, American Petroleum Institute (API) standards can also be used in addition to ASTM. In refineries and petrochemical plants, API 5L Gr.B piping may be used instead of ASTM A106 Gr.B for carbon steel piping. Additionally, API 5L grade X is commonly used for onshore pipelines. In the introduction to this chapter, the difference between piping and pipeline is explained. The main difference between API 5L Gr.B and ASTM A106 Gr.B is that A106 Gr.B is a seamless pipe, while API 5L piping may be either seamless or welded. As a matter of fact, API 5L steel pipes are derived from API specification 5L, and they are ideal for transportation of oil, gas, and water. It is important to define the product safety level (PSL) of API 5L pipes in the specification. PSL2 API 5L pipes are of higher quality than PSL1 API 5L pipes. In contrast to PSL1, PSL2 contains enhanced requirements, such as a mandatory notch toughness test, limited strength and chemical composition ranges, and a carbon equivalent for improving weldability. Moreover, PSL2 is used for piping larger than 4", so it cannot be used for threaded or plain ends.

2.2.7 Manufacturing Method

Another important piping parameter is the manufacturing method. There are two types of pipes: welded and seamless. A welded pipe is primarily composed of plates; these plates are rolled into a tube shape with the assistance of a plate bending machine or roller, and then the gap or seam is welded longitudinally. It is possible for piping smaller than 14" to be seamless. For seamless pipes, the joint efficiency is 1 or 100% because there is no seam weld. A billet of steel can be used to manufacture seamless pipes. By heating the billet in a furnace and using a piercer, referred to as a mandrel, along with rollers, it is possible to make a hole within the billet to form a pipe. As an alternative, molten steel can be extruded into a tubular shape to form seamless pipes. The process of making seamless pipes in large sizes can be challenging and can result in misalignment between the mandrel and billet, so a size limit may be specified for seamless pipes in the piping specification, such as a maximum of 12", 10", or 8". Welded pipe does not have any size restrictions. However, in the case of welded pipes that are larger than 36" in size, two plates are welded together instead of one, then bent and welded together. A welded pipe of smaller diameter consists of one plate that has been bent and welded together. In Figure 2.6, seamless pipes are shown without traces of weld seams. In Figure 2.7, a welded pipe is shown prior to welding the longitudinal seam of the weld. The longitudinal weld joint quality factor is based on ASME B31.3 Table 302.3.4 (see Table 2.1). Generally, it can range from 0.60 to 1.00, depending on the type of welding and the extent of the radiography test examination. A weld's joint efficiency is defined as the ratio of the weld joint strength to the strength of the base material. Ideally, the weld joint strength should equal the strength of the base material, resulting in a joint efficiency of 1.00 or 100%.

The furnace butt welding process refers to the method of welding a pipe. The pipe is heated in a furnace to a specific temperature. Welding is then performed on the longitudinal seam forge. Welding of this type is considered low quality, with weak joint efficiency. The electric fusion welded (EFW) pipe is a welded

FIGURE 2.7 Longitudinal weld seam between two sections of a bended pipe.

TABLE 2.1
Longitudinal Weld Joint Quality Factor for Piping according to ASME B31.3 Code, Table 302.3.4

No.	Type of Joint	Type of Seam	Examination	Weld Joint Efficiency (*E*)
1	Furnace buttweld	Straight	As required by the specification or code	0.6
2	Electrical resistance weld	Straight or spiral	As required by the specification or code	0.8
			Additional spot radiography	0.9
			Additional 100% radiography	1.00
3	Electrical fusion weld (single butt weld)	Straight or spiral	As required by the specification or code	0.8
			Additional spot radiography	0.9
			Additional 100% radiography	1.00
4	Electrical fusion weld (double butt weld)	Straight or spiral	As required by the specification or code	0.8
			Additional spot radiography	0.9
			Additional 100% radiography	1.00
5	Submerged arc weld, gas metal arc weld API 5L	Straight with one or two seams spiral	As required by the specification	0.95

pipe that is formed by rolling a plate and welding the seams together. Electric arc welding is a process in which metal is heated by electrodes (filling metal), and the electric arc is used to create the welding. EFW pipe is the most popular process for large-diameter pipes since it uses a longitudinal seam weld. Electric resistance welding is also known as ERW. ERW pipe is welded differently than EFW pipe and submerged arc pipe, as will be discussed later. We use the pressure welding method without filler metal. Welds are not filled with other components. In submerged arc welding (SAW), an electric arc is formed between a continuously fed electrode and the workpiece to be welded. A blanket of powdered flux surrounds and covers the arc and, when molten, provides electrical conduction between the electrode and the metal to be joined. In Table 302.3.4, the joint efficiency factor for a furnace butt weld is 0.6, making it unsuitable for use in oil and gas piping. In accordance with ASME B31.3, the joint efficiency of a furnace butt weld cannot be increased through the use of radiography examination. For seamless pipes, the joint efficiency is 1 or 100%, since there is no seam weld.

Therefore, in summary, the following parameters are required to be defined for a pipe in a piping specification:

Size, pipe thickness or schedule, length of pipe, pipe ending, method of pipe manufacturing, material grade, standard, and pipe ending standard. The pipe ending standard is applicable to buttwelded and threaded-end pipe connections.

Example 1) 1.5" SCH.XS, single random length, PE, seamless pipe, A106 Gr.B, ASME B 36.10

Example 2) 24" SCH.10S, single random length, BE, welded pipe, EFW, A358 Gr.304 CL.1, 100% radiography test (RT) on the welding, ASME B36.19M, ending per ASME B16.25

Example 3) ½" SCH 80, single random length, TE, seamless pipe, A106 Gr.B, ASME B36.10M, ASME B1.20.1

Table 2.2 provides the correct ASTM grades for different piping components, such as seamless and welded pipe, forged, cast, and wrought. Throughout this chapter, forged and wrought fittings, as well as casting, will be discussed in more detail.

Descriptions of the ASTM standards used in Table 2.2 are as follows:

Seamless Pipes

ASTM A106: Standard specification for seamless carbon steel pipe for high-temperature service

ASTM A333: Standard specification for seamless and welded steel pipe for low-temperature service and other applications with required notch toughness

TABLE 2.2
ASTM Material Grade Selection

Material Type	Seamless Pipe	Welded Pipe	Plate	Forge	Cast	Wrought (Note 1)	Bar (Note 2)
Carbon steel	A106 Gr. B	A672 C60/C65/C70 (CL22) *(Note 3)*	A515 Gr.65 or Gr70	A105	A216 WCB	A234 WPB	A105 bar
Low-temperature carbon steel	A333 Gr.6	A671 CC60/CC65 /CC70 (CL22) *(Note 3)*	A516 Gr.65 or Gr70	A350 LF2	A352 LCC	A420WPL6	A350 LF2 bar
Low alloy steel (1.25 Cr-0.5mo)	A335 Gr. P11	A691 1.25Cr CL22 *(Note 3)*	A387 Gr.11 CL2 *(Note4)*	A182 F11 CL2 *(Note4)*	A217 WC6	A234 Gr. WP11 CL2 *(Note 4)*	A182 F11 bar
Austenitic stainless steel	A312 TP304, 304L,316,316L etc.	A358 Gr.304, 304L, 316, 316L, etc.	A240 Gr.304/304L/ Gr.316/316L /321	A182 F.304/304L F.316/316L /321, etc.	A351 Gr. CF8/ CF8M/ CF8C	A403 WP304/316 304L/316/321 etc.	A479 S30400/30403 S31600/S31603 *(Note 5)*
22 Cr duplex	A790 S31803	A928 S31803	A240 S31803	A182 F51	A995 J92205	A815 S31803	A479 S31803 *(Note 5)*
25 Cr super duplex (Note 6)	A790 S32750/ S32760	A928 S32750/ S32760	A240 S32750/ S32760	A182 F53/A182 F55	A890 or 995 Gr.5A/A890 or A995 Gr.6A	A815 S32750/ S32760	A479 32750/ S32760
6MO	A312 S31254	A358 S31254	A240 S31254	A182 F44	ASTM A351 CK3-MCuN	A403 S31254	A479 S1254 *(Note 5)*
Inconel 625	B444 N06625	B705 N06625	B443 N06625	B564 N06625	A494 N06625	B366 N06625	B446 N06625

Copper-nickel (90/10)	B466 UNSC70600	B467 UNSC70600	B122 or B171 90Cu-10Ni	90Cu-10Ni	90Cu-10Ni	90Cu-10Ni	B151 90Cu-10Ni
Copper-nickel (70/30)	B466 UNSC70600	B467 UNSC70600	B122 or B171 70Cu-30Ni	70Cu-30Ni	70Cu-30Ni	70Cu-30Ni	B151 70Cu-30Ni
Titanium	B861 Gr.2	B862 Gr.2	B265 Gr. 2	B381 F2	B367 C2	B363 WPT2	B348 Gr.2

***Note 1*:** The process of manufacturing wrought fittings from pipe is explained in more detail in Chapter 2. The raw material (piping) used to make wrought fitting could be seamless or welded. The letter “S” designates a wrought fitting from a seamless pipe (e.g., A403 WP304-S), and “W” designates a wrought fitting from a welded pipe (e.g. A420 WP6-W); these letters are added at the end of wrought ASTM material grades.

***Note 2*:** The bodies of valves can be made from bar materials instead of being forged or cast to reduce manufacturing time. Bar materials can be used for other parts of the valve, such as seat, gland, etc. The usage of bar instead of forge for the body of a valve could be subject to mechanical tests, and such usage should be approved by the material department.

***Note 3*:** CL22 is selected for welded pipe in carbon, low-temperature, and low-alloy steels, meaning that PWHT is uniformly applied to the weld joint to release residual stress. Applying PWHT reduces the chance of stress cracking corrosion types by corrosive media such as hydrogen sulfide.

***Note 4*:** CL2 for plated, forged, and wrought low-alloy steel affects the mechanical strength of the material. CL1 has lower mechanical strength than CL2, and CL2 has lower mechanical strength than CL3 for the given materials.

***Note 5*:** A479 is a bar material used for stainless steel and 6MO. This material is used for the pressure-containing parts of valves such as the body and bonnet. Pressure-containing parts of valves are defined as components whose failure to function causes leakage. A276 is an alternative bar material for stainless steels, as is 6MO, which can be selected for non-pressure-containing parts of valves such as the gland and seat.

***Note 6*:** Two 25Cr super duplex grades are considered in the table: UNS S32750 and UNS 32760.

ASTM A335: Standard specification for seamless ferritic alloy-steel pipe for high-temperature service

ASTM A312: Standard specification for seamless, welded, and heavily cold-worked austenitic stainless-steel pipes

ASTM A790: Standard specification for seamless and welded ferritic/austenitic stainless-steel pipe

ASTM B444: Standard specification for nickel-chromium-molybdenum-columbium alloys (UNS N06625 and UNS N06852) and nickel-chromium-molybdenum-silicon alloy (UNS N06219) pipe and tube

ASTM B466: Standard specification for seamless copper-nickel pipe and tube

ASTM B861: Standard specification for titanium and titanium alloy seamless pipe

Welded Pipes

ASTM A672: Standard specification for electric-fusion-welded steel pipe for high-pressure service at moderate temperatures

ASTM A671: Standard specification for electric-fusion-welded steel pipe for atmospheric and lower temperatures

ASTM A691: Standard specification for carbon and alloy steel pipe, electric-fusion-welded for high-pressure service at high temperatures

ASTM A358: Standard specification for electric-fusion-welded austenitic chromium-nickel stainless steel pipe for high-temperature service and general applications

ASTM A928: Standard specification for ferritic/austenitic (duplex) stainless steel pipe electric fusion welded with addition of filler metal

ASTM B705: Standard specification for nickel-alloy (UNS N06625, N06219 and N08825) welded pipe

ASTM B467: Standard specification for welded copper-nickel pipe

ASTM B862: Standard specification for titanium and titanium alloy welded pipe

Plates

ASTM A515: Standard specification for pressure vessel plates, carbon steel, for intermediate- and higher-temperature service

ASTM A516: Standard specification for pressure vessel plates, carbon steel, for moderate- and lower-temperature service

ASTM A387: Standard specification for pressure vessel plates, alloy steel, chromium-molybdenum

ASTM A240: Standard specification for chromium and chromium-nickel stainless steel plate, sheet, and strip for pressure vessels and for general applications

ASTM B443: Standard specification for nickel-chromium-molybdenum-columbium alloy and nickel-chromium-molybdenum-silicon alloy plate, sheet, and strip

ASTM B122: Standard specification for copper-nickel-tin alloy, copper-nickel-zinc alloy (nickel silver), and copper-nickel alloy plate, sheet, strip, and rolled bar

ASTM B171: Standard specification for copper-alloy plate and sheet for pressure vessels, condensers, and heat exchangers

ASTM B265: Standard specification for titanium and titanium alloy strip, sheet, and plate

Forges

ASTM A105: Standard specification for carbon steel forgings for piping applications

ASTM A350: Standard specification for carbon and low-alloy steel forgings, requiring notch toughness testing for piping components

ASTM A182: Standard specification for forged or rolled alloy and stainless-steel pipe flanges, forged fittings, and valves and parts for high-temperature service

ASTM B564: Standard specification for nickel alloy forgings

ASTM B381: Standard specification for titanium and titanium alloy forgings

Castings

ASTM A216: Standard specification for steel castings, carbon, suitable for fusion welding, for high-temperature service

ASTM A352: Standard specification for steel castings, ferritic and martensitic, for pressure-containing parts, suitable for low-temperature service

ASTM A217: Standard specification for steel castings, martensitic stainless and alloy, for pressure-containing parts, suitable for high-temperature service

ASTM A351: Standard specification for castings, austenitic, for pressure-containing parts

ASTM A995: Standard specification for castings, austenitic-ferritic (duplex) stainless steel, for pressure-containing parts

ASTM A890: Standard specification for castings, iron-chromium-nickel-molybdenum corrosion-resistant, duplex (austenitic/ferritic) for general application

ASTM A494: Standard specification for castings, nickel and nickel alloy

Wrought

ASTM A234: Standard specification for piping fittings of wrought carbon steel and alloy steel for moderate and high temperature service

ASTM A420: Standard specification for piping fittings of wrought carbon steel and alloy steel for low-temperature service

ASTM A403: Standard specification for wrought austenitic stainless steel piping fittings

ASTM A815: Standard specification for wrought ferritic, ferritic/austenitic, and martensitic stainless steel piping fittings

ASTM B366: Standard specification for factory-made wrought nickel and nickel alloy fittings

Bars

ASTM A105: Standard specification for carbon steel forgings for piping applications

ASTM A276: Standard specification for stainless steel bars and shapes

ASTM A479: Standard specification for stainless steel bars and shapes for use in boilers and other pressure vessels

ASTM B446: Standard specification for nickel-chromium-molybdenum-columbium alloy (UNS N06625), nickel-chromium-molybdenum-silicon alloy (UNS N06219), and nickel-chromium-molybdenum-tungsten alloy (UNS N06650) rod and bar

ASTM B151: Standard specification for copper-nickel-zinc alloy (nickel silver) and copper-nickel rod and bar

ASTM B348: Standard specification for titanium and titanium alloy bars and billets

2.3 PIPE FITTINGS

A pipe fitting is a component that is used in piping systems in order to change the direction of the piping, to make a branch from a header, to reduce or expand piping size, or to blind the end of the pipe.

2.3.1 Fittings for Piping Route Changes

2.3.1.1 Elbows

An elbow is the most common way to change the direction of piping at two standard angles: 90° and 45°. An elbow with a 90° rotation angle is shown on the left and one with a 45° rotation angle on the right in Figure 2.8.

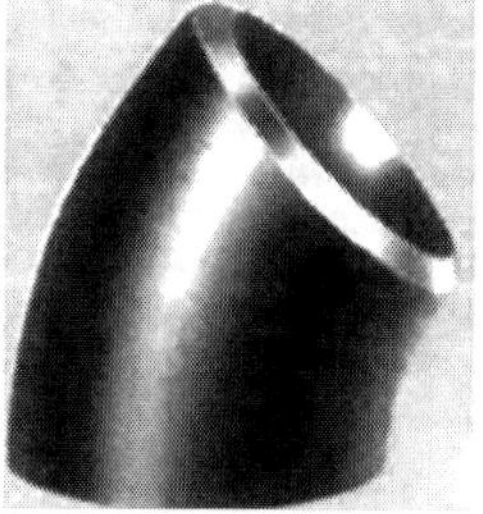

FIGURE 2.8 90° and 45° rotation angle elbows.

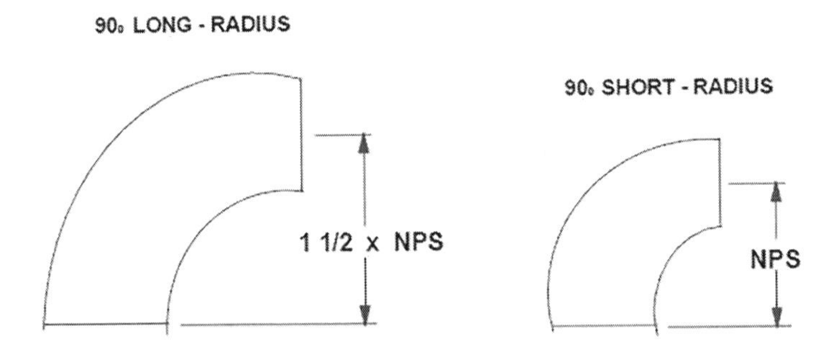

FIGURE 2.9 Long and short radius elbow comparison.

The radius of an elbow, whether it is a 45° or 90° angle, can be divided into two categories: short radius and long radius. A long radius elbow has a radius that is 1.5 times the diameter of the pipe or nominal pipe size, while a short radius elbow has a radius equal to the diameter of the pipe or NPS. It is shown in Figure 2.9 that a 90° elbow with a long radius and a short radius can be compared. A short radius elbow allows for a more compact design, but friction and pressure drop are high, so the present author does not recommend this type of elbow. The use of a short radius elbow may result in a significant pressure drop in the piping, requiring the use of a large pump or compressor in order to compensate.

2.3.1.2 Returns

For a 180° piping rotation, a component called a return, shown in Figure 2.10, should be selected.

Size, thickness, or schedule (for a wrought elbow with a bevel end) or pressure rating (for a forged elbow with threaded or plain ends), end, type of elbow (radius and angle of rotation), manufacturing type (forge or wrought), material grade, and standard are essential parameters for defining pipe fittings such as elbows.

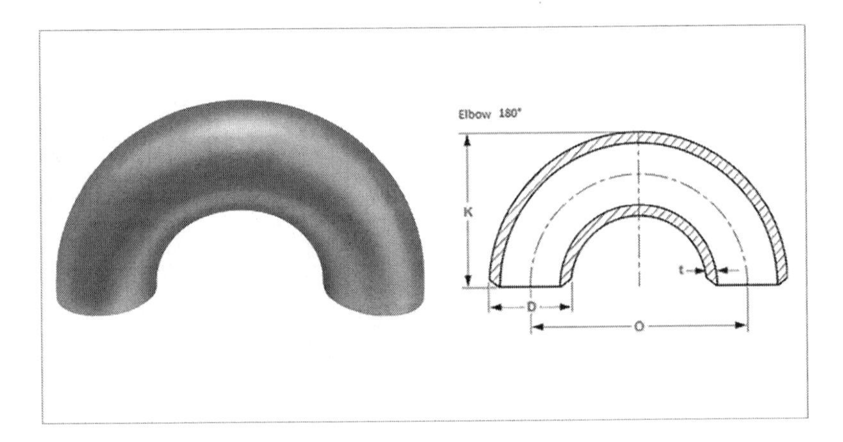

FIGURE 2.10 Pipe return.

There are two types of pipe fittings manufactured: wrought and forged. A standard known as ASME B16.9 applies to factory-made, wrought, buttwelded fittings. The specification specifies the overall dimensions, tolerances, ratings, testing, and marking of wrought fittings between nominal pipe sizes 12" and 48". Typically, fittings made in accordance with this standard are provided with butt-welded ends or bevel ends in accordance with ASME B16.25. Fittings made from raw materials such as pipe, plate, and forging are known as wrought fittings. Wrought fittings can be manufactured using a variety of methods, such as the hydraulic bulge method, hot extrusion, or other cold or hot forming methods. Hydraulic forming involves placing a piece of pipe in a hydraulic die, pouring liquid into the pipe, and allowing hydraulic pressure to push the pipe into the desired fitting shape. It should be noted that no welding is used in the wrought process. When the wrought fittings are buttwelded to the piping, they will have the same thickness as the mating or connected pipe. For example, a wrought butt weld elbow is connected to a 6" pipe with 10S schedule, which has a wall thickness of 3.4 mm. Therefore, the thickness of the wrought-made elbow should be 3.4 mm. Forging is another method of manufacturing piping fittings. Typically, forged fittings are used for small sizes, such as 4" and smaller. They are covered by ASME B16.11, "Forged Fittings, Socket-Welding and Threaded," which has threaded or socket weld ends. As illustrated in Figure 2.11, socket welding is a method of attaching piping by inserting the pipe into the recessed area of the fitting. The outer surface of the pipe is fillet welded to the fitting after the pipe has been inserted into the female area of the fitting. In the fitting, there is a gap between the pipe and the female end of the socket. As a result of possible thermal expansion of the piping during the fillet welding process, there is a 1.6-mm gap. Due to the possibility of corrosive fluid accumulating there, this gap is considered a weak point. Socket welding is not recommended for services that may create crevices and pitting corrosion, such as seawater.

As a result of forging steel, very strong fittings are produced. As soon as the material reaches molten temperatures, it is placed in the dies. Following the heating of the steel, the fittings are machined from the heated steel. As part of the ASME B16.11 standard, forged fittings are designated with pressure classes rather than wall thickness values. For threaded fitting connections, pressure classes of

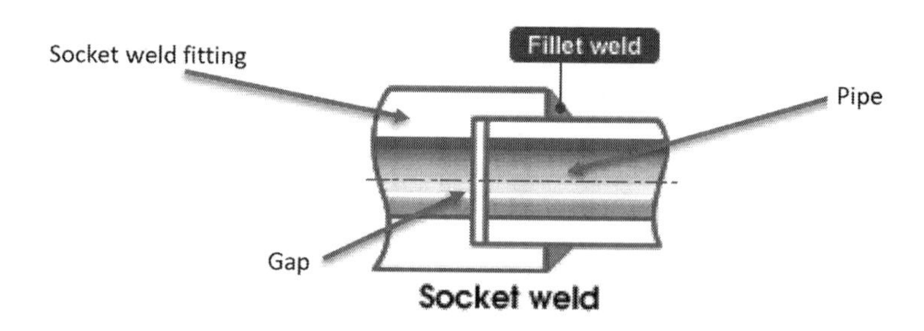

FIGURE 2.11 A socket weld between a pipe and socket weld–forged fitting.

TABLE 2.3
Correlation of the Fitting with Schedule Number or Piping Wall Thickness for Selection of Forged Fitting Rating per the ASME B16.11 Standard

Fitting Class	Type of Fitting	Schedule Number	Wall Designation
2000	Threaded	80	XS
3000	Threaded	160	-
6000	Threaded	-	XXS
3000	Socket-welding	80	XS
6000	Socket-welding	160	-
9000	Socket-welding	-	XXS

2000, 3000, and 6000 are used, and for socket weld–ended fittings, pressure classes of 3000, 6000, and 9000 are used. What are these numbers indicating? The pressure class of a fitting indicates how much pressure it can withstand in pounds per square inch. For instance, a threaded or socket end fitting in pressure class 3000 can withstand maximum pressures of 3,000 pounds per square inch. Threaded fittings are not recommended for pressure ratings of 2,000 due to their insufficient robustness. The main question is, how can the fittings be selected based on the thickness of the piping wall? In Table 2.3, which is taken from ASME B16.11, the fitting pressure class is correlated with the thickness of the piping wall or the schedule number. Based on pressure class ratings, the ASME B16.11 standard specifies the dimensions of fittings, including wall thickness.

2.3.1.3 Bends

In spite of the fact that a long radius elbow is selected, the pressure drop between the piping and the elbow may exceed the limits indicated by the process engineers. It is possible to experience a high pressure drop due to dense flow or abrasive fluid services, for example. In order to minimize the pressure drop in the piping system, pipe bends may be preferred over elbows on the discharge piping from compressors or pressure safety valves. Bends are preferred in such a situation, since they have a longer radius than elbows. As a result of the larger radius of a bend, this component is less compact than an elbow, but it is capable of reducing pressure drop. A bend may have a radius three times the diameter of the pipe or NPS or five times the diameter of the pipe or NPS. A bend is considered a special item in the piping industry. It is possible to make bends in different ways, such as by cold bending or hot bending. Factory-made induction bends are very common, and they are manufactured in compliance with ASME B16.49. Unlike construction bends, which are made in yards, factory-made bends are manufactured by a manufacturer. By applying heat locally with frequent electrical pulses, induction bending is a controlled method for bending pipes. Upon reaching the correct bending temperature, the pipe is slowly moved through the induction coil while bending force is applied to bend it. Since bends are made using the wrought

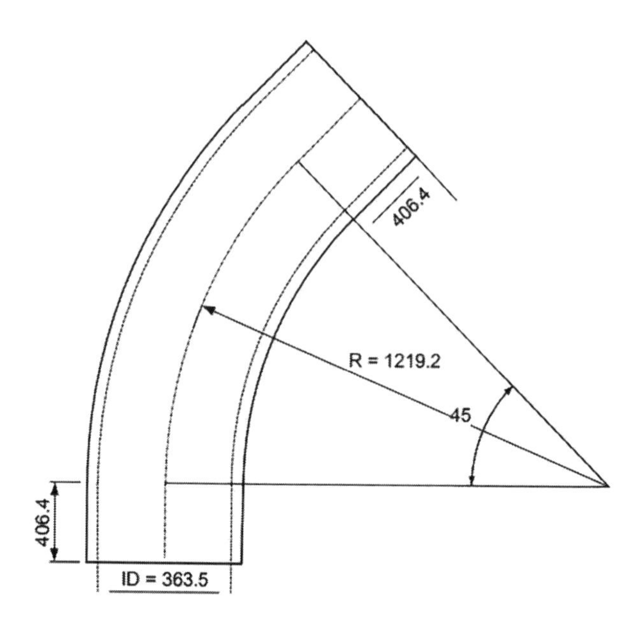

FIGURE 2.12 A 16", 3D, 45° bend.

process, all of the parameters described for wrought elbows apply to bends, with the exception of the design standard. It is useful to know that a bend may have a rotation angle of 45° or 90°. Due to the fact that bends are specially made items, other angles, such as 30°, may be required for certain projects, depending on the layout of the pipes. A 16-inch 3D bend with a 45° rotation is illustrated in Figure 2.12. The ovality or roundness of the bend is another significant parameter that should be considered in the specification related to bends. The ovality of a circle is a measure of its deviation from circularity. ASME B16.49 allows a maximum ovality of 3% of the nominal outside diameter of the mating pipe within the bend and 1% of the nominal outside diameter of the mating pipe at the welding end. At both ends of welded bends, less ovality is tolerated. The purchaser and manufacturer may, however, agree on other values for the ovality of the bend.

2.3.1.4 Miters (Miter Bends)

A miter is defined in ASME B31.3, process piping code, as "two or more straight sections of pipe matched and joined in a plane bisecting the angle of junction in order to produce a change in direction." Miter fittings are not standard fittings; they are manufactured from pipe. Due to the limitations and disadvantages of miter bends, many end users and contractors do not allow their use. Following is a summary of the limitations of a miter bend:

1. Miter bends are made up of numerous pieces that are welded together, as opposed to a one-piece integrated elbow or bend, and therefore have a low degree of strength.

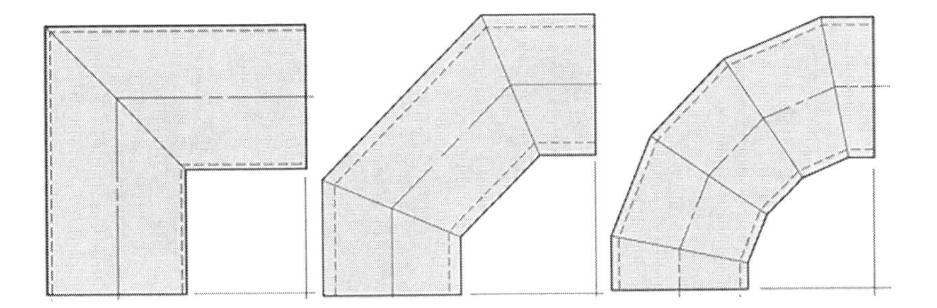

FIGURE 2.13 Miter bends with two, three, and five pieces of pipe.

2. As a result of the high pressure drop inside a miter bend coupled with high flow turbulence, erosion may be more likely to occur.
3. It is possible that erosion and corrosion will be exacerbated by the presence of corrosive fluid inside the pipe, which may increase the corrosion rate in the miter, especially on the weld joints.
4. In order to make a miter bend, a skilled pipe fitter is required.

In contrast, miter bends are characterized by their low cost and ease of fabrication in the fabrication yard. As a result, there is typically no need to order a miter from a piping manufacturer. Instead, piping pieces are cut and welded together in order to create a miter. In a miter bend, each two-pipe joint is buttwelded together. Figure 2.13 illustrates how up to five pieces of pipe can be used to make a miter bend. In addition to being more expensive and more difficult to fabricate, miters with more cuts require more welding points, but they also reduce pressure drop and provide smoother flow. There is a 90° rotation of piping using all pieces of the miter bends shown in the figure. By using a miter bend, it is possible to achieve a 45° rotation of the piping. Miter bends may also have long or short radius values equal to D (pipe diameter), 2D, 3D, 5D, 5D, 8D, 10D, or custom shapes.

2.3.2 Fittings for Branching

A second application of fittings is the construction of branch connections. A branch connection should be selected based on the size limitations of the branch fittings, the ASME B31.3 process piping code requirements, and the cost of the material and fabrication. It is generally accepted that four types of branching components or approaches are used in piping engineering: tees, laterals, olets, and pipe-to-pipe or stub-in connections. It is possible to reinforce pipe-to-pipe or stub-in connections with a pad. The various branches are described as follows.

2.3.2.1 Tees

For most applications, a tee is the strongest and safest branch connection. As a result of their resistance to fatigue and vibration, tees are always used in piping

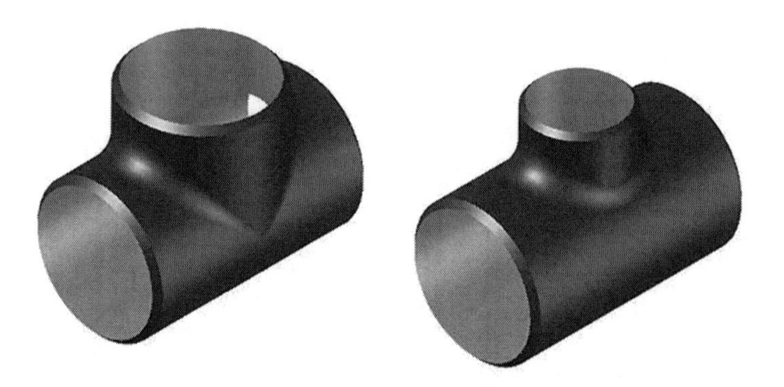

FIGURE 2.14 An equal tee (left) and a reducing tee (right).

systems in which the header and branch are of equal size. A tee is primarily used to make 90° branches from a pipe header. Tees can be divided into two types: equal or straight tees and reduced tees. If the diameter of the pipe header and the diameter of the branch are equal, a straight or equal tee should be used. Reduced tees are used when the branch is smaller than the header. For the purpose of defining a straight tee with equal header and branch sizes, only one size can be used. It is possible to introduce a reduced tee size by two or three. For example, if you want to connect a 3" pipe to a 2" branch, you can introduce the size as 3" × 2" or 3" × 3" × 2". Reducing tees have a major limitation in that they are not included in ASME B16.9, the standard for wrought fittings, when there is a substantial difference between the header and the branch. If the header is twice as large as the branch, it is usually possible to use a reduced tee. For example, a reduced tee can connect an 8" header to a 6" or 4" branch size, but an 8" header to a 3" branch does not exist in the standard, so an olet should be used in this case, as described later in this chapter. An equal tee is shown on the left, and a reducing tee is shown on the right in Figure 2.14.

The following parameters must be considered when defining a tee: size (large and small, or just one size for an equal tee), thickness (typically two thickness values, one for each of the large and small ends of a reduced tee and one thickness value for equal tees), end connection(s), type of tee, method of manufacture, whether it is forged or wrought (if the tee is wrought, please specify whether the pipe is seamless or welded), material grade, standard.

Example 1) 3" × 2", SCH 40 × SCH 80, BW, reducing tee, wrought from seamless pipe, ASTM A234 WPB, ASME B16.9.

Example 2) 1.5", CL3000, PE, equal tee, forged, ASTM A105, ASME B16.11

2.3.2.2 Olets

Essentially, an olet fitting connects a larger pipe to a smaller one by providing an outlet. As discussed previously, when a pipe's header and branch have a significant size difference, a tee cannot be used because the relevant standard limits the sizes. On the other hand, olets are proposed based on the author's experience and the industry's norm. Weldolets, sweepolets, sockolets, threadolets, elbolets, and flangeolets are all examples of reinforced branches. The MSS SP 97 standard covers specs for olets, including dimensions, finishing, tolerances, testing methods, marking, and material grade. Forging is a common method of fabrication for olets. Weldolets are 90° branch connections that are welded to pipes from one side; on the branch side, a bevel is attached to another piping component to secure the connection. In Figure 2.15, the weld between the weldolet and the header is a fillet weld. When two joints are perpendicular or have an angle, fillet welding is used to join them together.

Sweepolets (see Figure 2.16) are commonly used in high-pressure piping applications, such as subsea piping. Also, there are instances in which a sweepolet could be preferred over an olet in large pipe branches such as 10" and above. The

FIGURE 2.15 A weldolet welded to a pipe header.

(Courtesy: Shutterstock)

FIGURE 2.16 A sweepolet.

present author has worked on an onshore project in which a sweepolet was used for a 10" branch in connections of 24", 26", 28", and 30" to 56" headers. Designed for low stress and long fatigue life, Sweepolets provide high strength with a low stress identification factor.

Another type of olet with a socket weld outlet is the sockolet (see Figure 2.17). In fact, the main difference between a weldolet and a sockolet lies in the end connection to the branch; a weldolet has a bevel end for connecting to the branch by buttweld, whereas a sockolet has a female end for connecting to the branch by socket weld. In addition, a sockolet that has a socket weld connection on one side is typically used for small pipe sizes from ½" to 1 ½" (less than 2"). In this chapter, we have explained that socket weld connections are not recommended for use in corrosive environments due to the possibility of crevice corrosion.

As illustrated in Figure 2.18, a threadolet is another type of olet with a female thread outlet. Generally, threadolets are formed by filleting the pipe from one side

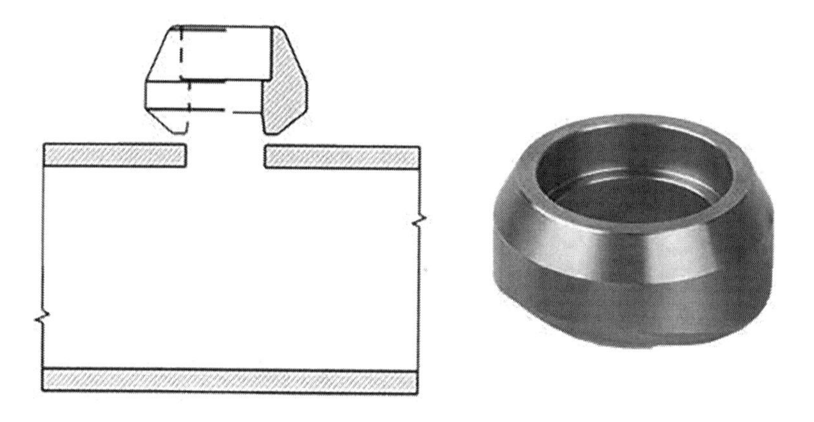

FIGURE 2.17 A sockolet.

FIGURE 2.18 A threadolet.

and connecting it to a male threaded pipe or nipple on the other side. In piping, a nipple refers to a short length of pipe, 100 or 150 mm, with male threads on both ends. A threadolet is used for small piping branch connections from ½" to 1½" (less than 2"); again, threaded end connections are not recommended for corrosive services and severe cycling conditions.

In general, olets offer three advantages over tees: less welding, lower space requirements, and faster fabrication.

2.3.2.3 Pipe-to-Pipe or Stub-In

Pipe branching may also be achieved by drilling a hole in the pipe header equal to the diameter of the pipe branch and fitting and welding the pipe header and branch together. The pipe-to-pipe method of branching, or stub-in, is more cost effective and less robust than both tees and olets. The lower cost of a pipe-to-pipe connection can be attributed to two main reasons. The first is that it eliminates the need to purchase a pipe fitting, and the second is that it is simple to construct by applying welding to the connection. For utility services such as water, pipe-to-pipe connections can be used in the onshore oil and gas industry. It is not recommended to use pipe-to-pipe connections in the offshore sector or for applications involving high pressures or aggressive fluids. Fluids that are aggressive are those that are flammable, toxic, or corrosive. The pipe-to-pipe connection may require a reinforcement pad (see Figure 2.19), which surrounds the branch and adds strength. Cutting a hole in the pipe header for the purpose of making a pipe-to-pipe or stub-in connection weakens the joints around the hole and concentrates the stress. If the thickness of the pipe header is insufficient to sustain the pressure, a reinforcement pad should be used to compensate for the thickness of the pipe header. Reinforcement pads are made of the same material as run pipes, and their wall thickness is the same as that of run pipes.

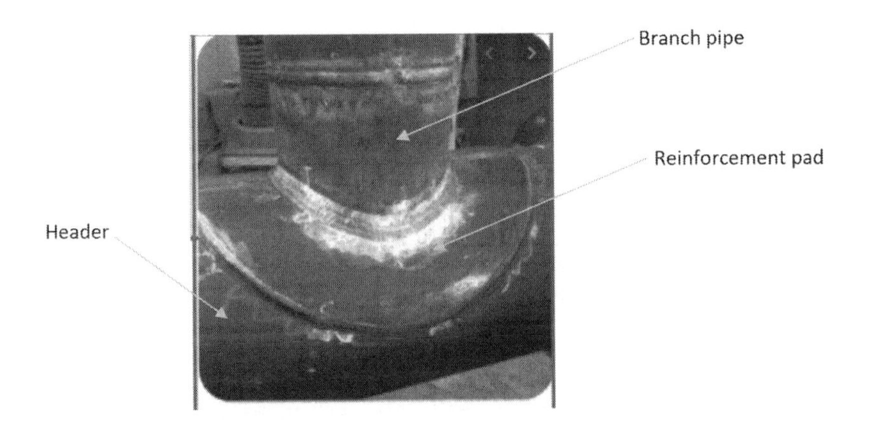

FIGURE 2.19 A reinforcement pad welded around a pipe-to-pipe connection.

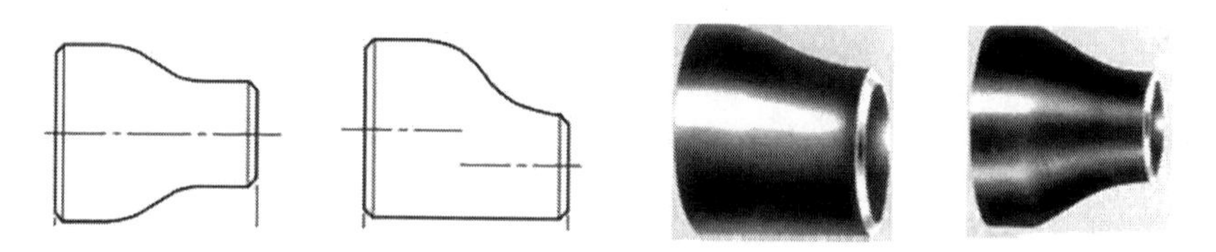

FIGURE 2.20 Eccentric and concentric reducers.

2.3.3 Fittings for Pipe Size Changes

2.3.3.1 Reducers

Reducers are components of piping systems that change the diameter of the pipe from a larger to a smaller one. Reducers are used to change the size of pipes in order to meet flow requirements or to accommodate existing piping. As shown in Figure 2.20, there are two types of reducers: concentric and eccentric. In concentric reducers, as shown on both the right and left sides of Figure 2.20, both ends of the reducer remain on the same axis. As a result, the centerline of the pipe remains at the same elevation, but the bottom of the pipe would change as a result of using a concentric reducer. As shown in Figure 2.20, eccentric reducers do not have their centers on the same axis; however, the bottom of the pipe remains the same elevation. An eccentric reducer with a flat bottom has the advantage that the bottom of the pipe and the length of the pipe support do not have to be altered. Eccentric reducers also have the advantage of being able to be used before or on the suction inlet of a pump mounted on a horizontal pipe. In order to install an eccentric reducer before the pump, the flat part of the reducer should face upwards. By utilizing this configuration, the top of the reducer is prevented from forming an air pocket. As an alternative, if a concentric reducer is placed at the suction of the pump, an air pocket may form and bubbles will be introduced, resulting in cavitation and eventual failure of the pump. A pump is a device that increases the head pressure of liquid services, so the presence of air or vapor in the pump is undesirable. It is actually the bubbles that are subjected to higher pressures inside the pump that implode on the surface of the impeller and cause irregularities, known as pits, which are forms of erosion corrosion. Pump impellers are rotating disks in centrifugal pumps that transfer energy from the motor to the fluid being pumped.

2.3.3.2 Swages

The swage, also known as a swage nipple or reducing nipple, is a type of forged fitting that connects two pipes of different sizes with threaded or plain ends. As shown in Figure 2.21, swage nipples are long enough to provide threaded ends on both sides. In order to make a buttweld connection, a swage may have a bevel end. Furthermore, it can have two different ends, such as one end threaded and one end beveled, one end threaded and one end plain, and so on. Figure 2.21 illustrates how a swage nipple may be concentric or eccentric, similar to a reducer.

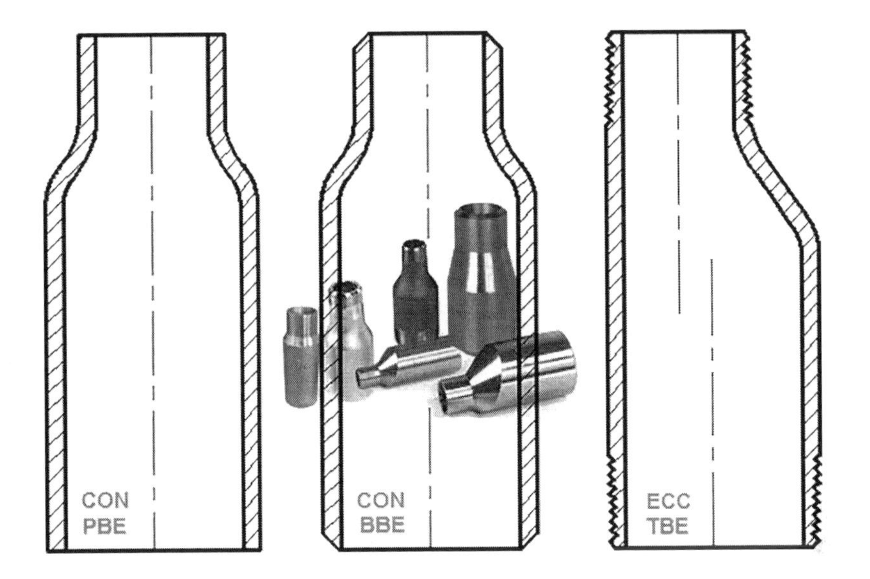

FIGURE 2.21 Swage nipples.

Swage nipples between 12" and 12" are covered by the MSS SP95 swage nipple standard.

2.3.4 Fittings for Pipe Termination or Blinding

Normally, pipe ends are not connected to any equipment and are blinded or capped with either a wrought bevel or a forged socket weld cap (applicable to sizes up to 1.5"), as shown in Figure 2.22.

Pipe plugs are another type of fitting used to close the ends of various small pipes and tubes, including those used to transport hydraulic fluid and pneumatic air. Plugs have a male thread at the end and are used to close the ends of pipes and tubes that have female threads. Tubes are hollow, elongated cylinders that are specially designed for the conveyance of fluids. Figure 2.23 illustrates a plug with

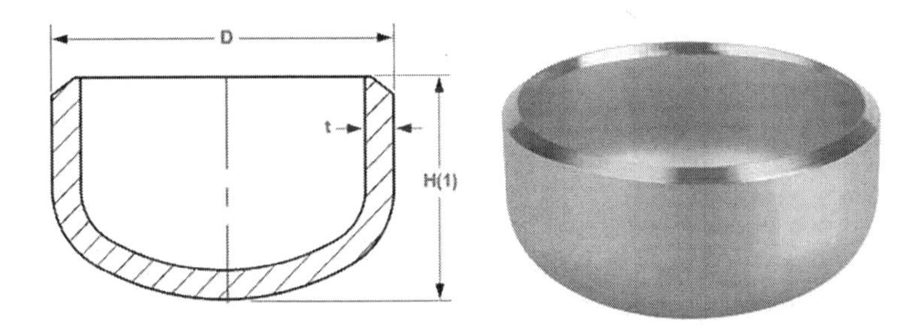

FIGURE 2.22 Wrought cap.

FIGURE 2.23 A plug with a square head.

a square head that has a threaded end. Plugs typically have a threaded end and are typically used in sizes of 1.5" and smaller.

2.4 FLANGES

As a component of a piping system, a flange connects pipes, valves, pumps, and other equipment. As far as maintenance is concerned, flange connections are preferred over welding due to their ability to be disassembled by removing the bolts and nuts connecting them to the pipe. As a result, when piping joints are welded together, it is difficult to perform maintenance on them. Despite the fact that many engineers perceive flange joints in piping systems as simple, the science related to flange joints and their sealing is in fact quite complex. Oil and gas production fields, refineries, and petrochemical plants require proper design, selection, and assembly of flange joints. As explained in this chapter, it is important to select the proper type and material of flanges, seals and gaskets, as well as bolting. To ensure accurate and leakage-free assembly of flange joints, a fitting approach and fitter skill are essential. As illustrated in Figure 2.24, flange connections consist of two mating flanges, a gasket, and bolting, referred to as nuts and bolts.

As explained in this chapter, each flange component must be designed, selected, and installed correctly in order to ensure safe and reliable long-term tightness and operation of the system. For the joint to function and operate properly in the piping system, all of the component parts of the assembled bolted flange joints are essential. The failure of a flange connection to achieve the required tightness can result in many negative consequences, such as lost production, environmental pollution, and injuries to personnel. Generally, sealed flange joints are not absolutely tight. It is inevitable that some leakage will occur from flange joints, as they are demountable. Flange connections can, however, be tightened to prevent leakage and reduce emissions. The following are different types of flanges.

FIGURE 2.24 A flange connection including mating flanges, a gasket, bolts and nuts.

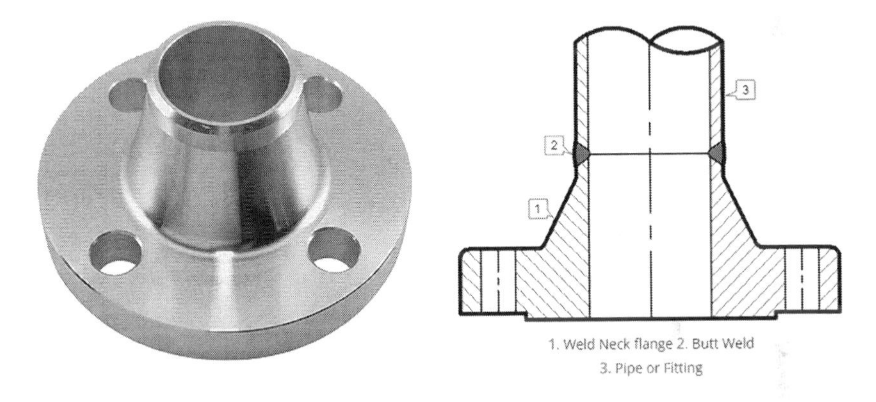

FIGURE 2.25 A weld neck flange.

2.4.1 Weld Neck Flanges

Weld neck flanges (WNFs) are robust and are suitable for high-pressure applications of pressure class 600 (approximately 100 bars) and higher and hazardous services that are toxic and/or flammable. In addition, WNFs are suitable for services that experience fluctuations in pressure and temperature. Nevertheless, WNFs can be used in the industry for low-pressure classes and non-aggressive fluids. The long and tapered hub of a WNF can easily be identified, as shown in Figure 2.25. The WNF (Item #1) is connected to a pipe or fitting (Item #3) by a buttweld on the right side of the figure.

2.4.2 Socket Weld Flanges

For small piping sizes up to 2", socket weld flanges are commonly used. As illustrated in Figure 2.26, the pipe (Item #3) is inserted into the socket end of the

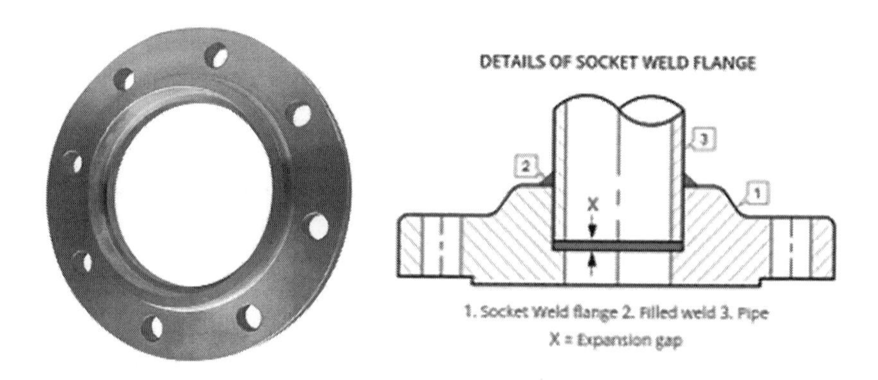

FIGURE 2.26 A socket weld flange.

flange (Item #1), and a fillet weld is applied to the top of the flange (Item #2). An important consideration is the gap between the flange and the pipe or fitting. As a general rule, this distance is approximately 1.6 mm. Its purpose is to provide a space or clearance for the pipe to expand as a result of the increased welding temperature. As a result of this gap, socket weld flanges are considered weak, since crevice corrosion may occur in piping that is exposed to corrosive services, such as sea water.

2.4.3 Slip-on Flanges

As shown in Figure 2.27, a slip-on flange (Item #1) is a type of flange that slides over a pipe (Item #4); thus its internal diameter is slightly greater than the pipe's. There are two fillet welds applied to the outside and inside of the flange, as shown in Items #2 and #3. In comparison to a WNF, slip-on flanges have lower strength and integrity under internal pressures and loads, including fatigue. This author recommends using slip-on flanges only for carbon steel materials and non-corrosive services with a maximum pressure class or rating of 300 equal

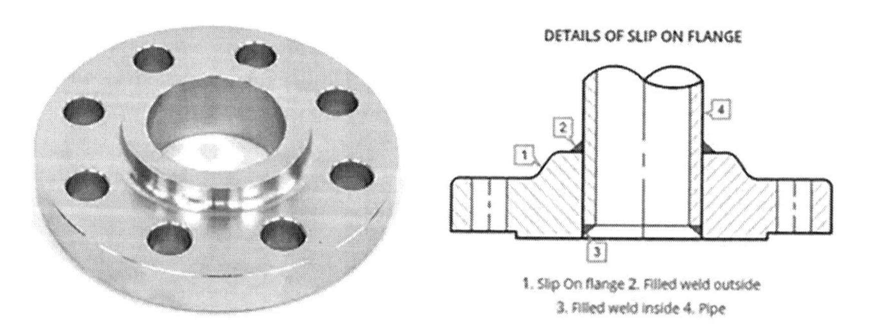

FIGURE 2.27 A slip-on flange.

to 50 bar. In corrosive environments, slip-on flanges should not be used. It is necessary for the client or end user of the plant to agree to the use of a slip-on flange by the contractor company. However, given all of its disadvantages, why is a slip-on flange sometimes preferred over a WNF? This is because the initial cost of a slip-on flange is lower than that of a WNF. Since a slip-on flange consists of two parts, it is called a loose flange. Another type of loose flange is a lap joint flange, which is discussed in the following subsection. For slip-on and socket-weld flanges, the height or size of the fillet weld should be the lesser of the flange thickness or 6 mm.

2.4.4 Lap Joint Flanges

Lap joint flanges are loose flanges used in conjunction with stub ends as shown in Figure 2.28. There are similarities between a slip-on flange and a lap joint flange, as both are types of loose flanges. It is important to note that the main similarity between these two flanges is that either a pipe or a stub end is seated inside both, and they both slip over the pipe or stub end. A slip-on flange as well as a lap joint flange can also be used in order to reduce costs. An offshore project experienced the use of a lap joint flange in corrosive seawater service as an example. Since the piping material was titanium, the stub end in contact with the fluid was also made of titanium. To reduce costs, the lap joint flange was made of stainless steel 316 rather than titanium. As one of the differences between a lap joint and a slip-on flange, the lap joint has a stub end that is buttwelded to the pipe from one end, whereas the slip-on flange has a fillet end that connects to the pipe. There is a low-pressure handling capacity of a lap joint flange, which is equal to 20 bars at maximum pressure class 150. As a lap joint flange has a low fatigue life, this type of flange is not recommended for use with equipment nozzles where vibrations and loads are applied to it.

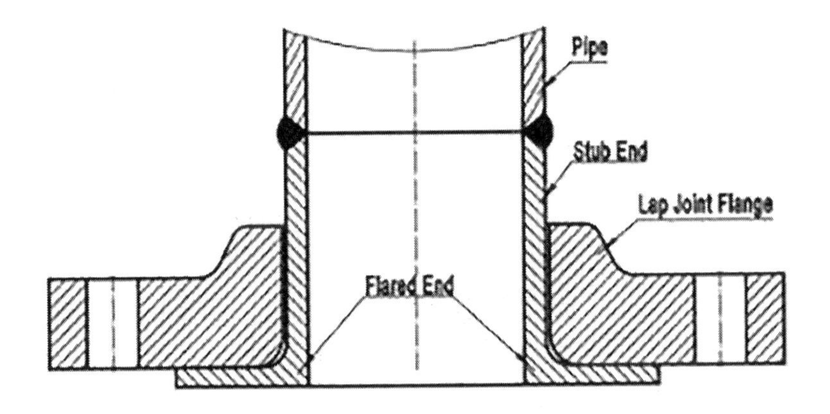

FIGURE 2.28 A lap joint flange.

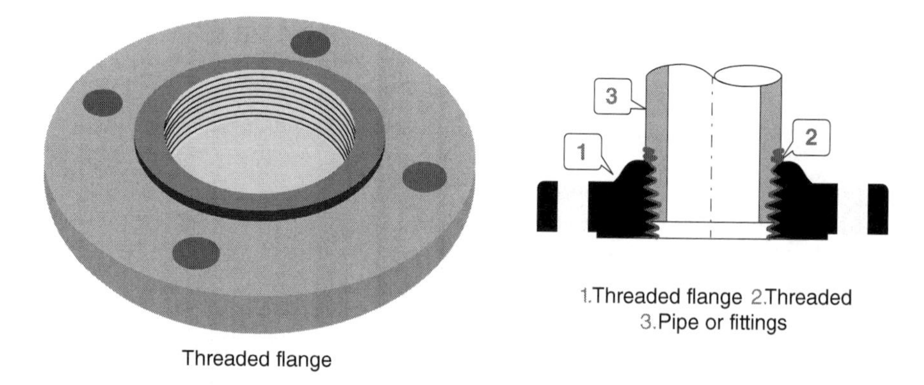

FIGURE 2.29 A threaded flange.

2.4.5 Threaded Flanges

The right side of Figure 2.29 illustrates a threaded flange (Item #1) connected to a male threaded pipe or fitting (Item #3) through threads (Item #2). In order to improve the load resistance of the threaded connection, a seal weld may be applied, which is a type of fillet weld. The use of threaded pipes is typically limited to small sizes, typically 4" and less, in applications where welding cannot be performed. In addition to the size limit, threaded joints and threaded flanges should not be used for corrosive, toxic, flammable, or high-pressure services.

Compared to both buttweld and socket weld connections, threaded connections are relatively weak; they can be loosened with vibration and any other type of load, resulting in leaks over time. Furthermore, the threads are susceptible to corrosion. Therefore, threaded flanges are an economical choice for non-corrosive, non-flammable services such as water and air.

2.4.6 Blind Flanges

A blind flange (also known as a closure plate flange) is used to terminate a pipe at the end of a piping system. The flange does not have a central hole (bore), so there is no flow through it. It may be necessary to use a blind flange in order to isolate a pipe, valve, or pressure vessel. Figure 2.30 illustrates an example of a blind flange blinding a WNF flange (Item #4) on the right side.

2.4.7 Orifice Flanges

The orifice flange is a type of piping flange that is used to measure the flow rate inside the pipe in conjunction with an orifice plate. There are typically two pressure tappings on each side of the flange, located opposite each other, and they are machined into the flange. Flow rates can be calculated by connecting pressure tappings to pressure gauges that measure pressure drops across orifice plates.

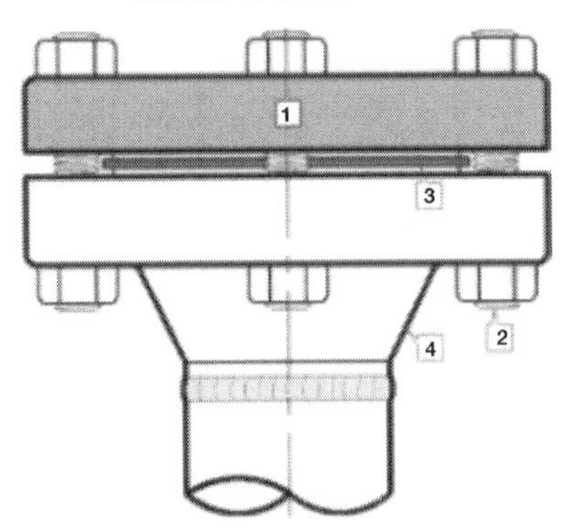

FIGURE 2.30 A blind flange.

FIGURE 2.31 An orifice plate between orifice flanges.

(Courtesy: Shutterstock)

By reducing pressure and restricting flow, an orifice plate measures flow rate. In Chapter 1, you will find more information about orifice plates. Figure 2.31 illustrates an orifice plate installed between two orifice flanges

2.4.8 Reducing/Expanding Flanges

A reducing flange (see Figure 2.32) is a type of flange that usually has a larger flange connection than the bore (e.g. 12" flange × 10" bore). For instance, a mixer of 12" size was purchased instead of a 10" mixer, and it should be connected to a 10" line. As a result, a reducing flange of 12" with a bore of 10" should be selected to connect the flange to the line.

Expander flanges are used to increase the pipe size by combining the size of the expander with a weld neck flange.

FIGURE 2.32 A reducing flange (larger flange compared to the piping connector).

Note 1: Flange faces are defined as the surface area on which the gasket is located. Flange faces can be smooth or serrated. There are five types of flange faces: flat face, raised face, ring type joint (RTJ), male and female, and tongue and groove. Various types of flange faces are defined in flange standards such as ASME B16.5 and ASME B16.47. In the case of metallic gaskets with metal facings, a smooth finish is recommended. A serrated finish can have a roughness between 125 and 500 microinches (μinch); it is suitable for non-metallic gaskets and semi-metallic gasket types such as spiral wound.

Note 2: For process and utility pipework, the ASME B16.5 and ASME B16.47 standards are commonly used. In accordance with ASME B16.5, pipe flanges and flanged fittings up to 24" in diameter are rated for pressure in the following pressure classes: 150, 300, 600, 900, 1500. Nevertheless, ASME B16.5 covers flanges up to and including 12" for a pressure class of 2500. Flanges and flanged fittings are discussed in this standard in terms of pressure ratings, dimensions, tolerances, materials, marking, and testing. ASME B16.5 does not include large-diameter flanges over 24"; therefore, ASME B16.47 is used for pressure rating, dimensions, tolerances, materials, marking, and testing of flanges from 26" to 60". There are two series of flanges in ASME B16.47: A and B. In accordance with ASME B16.47, Series A flanges are the same as MSS SP44 flanges. The Manufacturers Standardization Society of the Valve and Fittings Industry (MSS) is an organization that develops standards for fittings, flanges and valves, and SP stands for standard practices; MSS SP44 addresses steel pipe flanges. The American Petroleum Institute (API) 605, which addresses large carbon steel flanges, has been replaced by ASME B16.47 Series B. ASME B16.47 Series A flanges are typically thicker, heavier, and stronger than those in Series B. Due to their robust design, Series A flanges can withstand a greater amount of external loading than Series B flanges. Further, Series A flanges have larger bolt holes and bolt circle diameters than Series B, allowing fewer bolts to be used.

2.5 GASKETS

Metallic gaskets for pipe flanges are covered by ASME B16.20, and non-metallic flat gaskets for pipe flanges are covered by ASME B16.21. Several factors contribute to the failure and malfunction of gaskets, including improper selection of the gasket type, poor selection of the gasket material, insufficient lubrication, uneven surfaces, too much hardness or softness of the gasket, and inadequate bolt torque on the mating flanges between which the gasket is installed. In designing and selecting gaskets, several factors must be considered, including compatibility with the fluid, internal piping and flange pressure and temperature, vibration and cyclic loads in the piping system, the material of the flange, the integrity of the gasket against the flange face, and, last, economy. A gasket should be made of a softer material than the flanges in order to provide adequate sealing and be able to be deformed. Tightening the bolts causes the gasket materials to fill in the irregularities of the flange face, resulting in a tight seal. Additionally, gaskets should be made from materials that will not corrode when exposed to internal fluids in the pipe. This section discusses three types of gaskets: non-metallic flat gaskets, semi-metallic (spiral wound) gaskets, and RTJ gaskets.

2.5.1 Flat Gaskets

Flat gaskets made of non-metallic materials include aramid fibers, glass fibers, Teflon (PTFE), graphite, or elastomers. The ASME B16.21 standard covers flat gaskets. In addition to having a low sealing stress, non-metallic flat gaskets can also be deformed easily in order to accommodate irregularities in the flange face. Rubber is a type of plastic elastomer, and Figure 2.33 illustrates a flat gasket made of rubber. As a rule, non-metallic flat gaskets are used for low pressure classes, such as CL150 equal to 20 bar, as well as non-aggressive and non-process services, such as air and water. As an example, these gaskets could be 1 or 2 mm thick. Materials containing asbestos, which is a naturally occurring fibrous silicate mineral, should not be used for flat gaskets due to their potential health hazards. Asbestos is most commonly introduced into the body through the respiratory system. Considering that asbestos fibers are extremely difficult to destroy, asbestos entering the human body can cause lung disease and cancer. As a result, compressed asbestos fiber (CAF) gaskets should be replaced with non-asbestos fiber (NAF) gaskets.

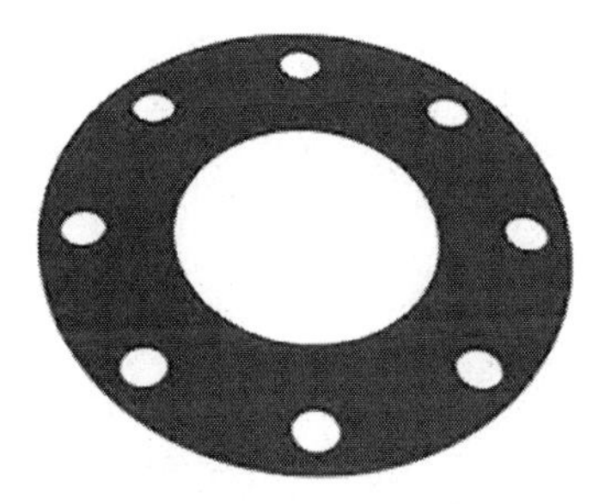

FIGURE 2.33 Flat rubber gasket.

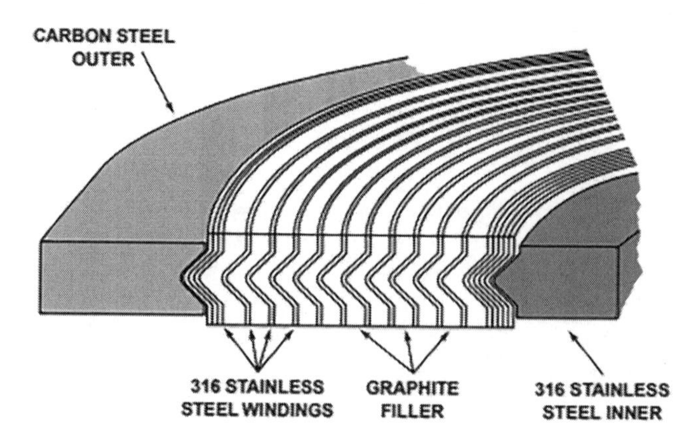

FIGURE 2.34 Spiral wound gasket.

2.5.2 Spiral Wound Gaskets

Gaskets that are spiral wound are semi-metallic gaskets that are made from both metallic and non-metallic materials. A spiral wound gasket is one of the most common types of semi-metallic gaskets, as illustrated in Figure 2.34. The spiral wound gasket is composed of three parts: the inner ring made of corrosion-resistant alloy, such as stainless steel 316, and the outer ring made of nickel alloy, such as Inconel 625, to increase its strength. The inner ring may be omitted in some cases, such as small gaskets. It is normal for the inner ring to have a thickness between 2.97 and 3.33 mm. Additionally, there is an outer ring, which is also a metallic component that may be constructed from carbon steel in onshore plants; this part, also known as a centering ring, is necessary for proper alignment of the gasket. Centering rings typically have the same thickness as inner rings, ranging from 2.97 to 3.33 mm. Finally, the winding and filler are located between the inner and outer rings and provide gasket sealing. Typically, this part is made of steel (such as stainless steel 316L) and graphite filler and measures 4.5 mm in thickness. Since spiral wound gaskets contain graphite filler, extra care should be taken when handling them. When spiral wound gaskets are in operation, large scratches on the graphite filler can cause leaks. Gaskets should not be used if the graphite filler has been removed due to a scratch, as the area where the graphite has been removed will leak. Spiral wound gaskets are covered by the ASME B16.20 standard. It may be restricted to place spiral wound gaskets between two weld neck flanges when the flanges are in pressure class 150. In contrast to a weld neck flange, spiral wound gaskets require a higher seating load, high-strength bolting, and proper bolting up techniques.

2.5.3 Metal Jacket Gaskets

A metal jacket gasket is another type of semi-metallic gasket. According to its name, it consists of a metallic outer shell and a non-metallic compressed filler

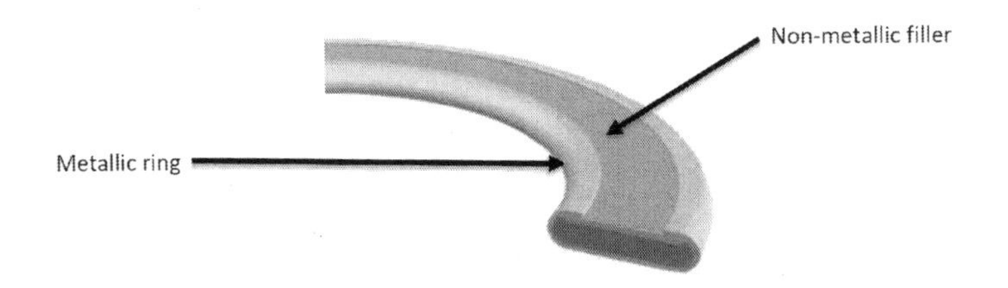

FIGURE 2.35 A metal jacket gasket.

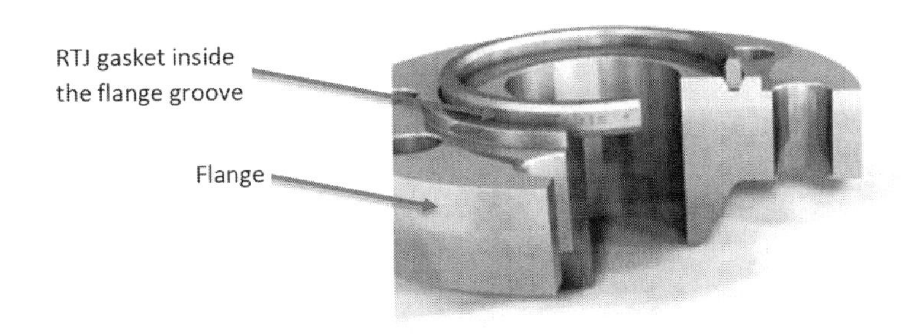

FIGURE 2.36 RTJ gasket inside the groove of a flange.

inside (see Figure 2.35). A filler material provides resilience to the gasket, while its metallic component provides resistance to pressure, temperature, and corrosion. There are a variety of metallic materials available, such as nickel alloy and stainless steel, while non-metallic materials include graphite, PTFE, and ceramics. Metal jacket gaskets are not common in piping systems in the oil and gas industry.

2.5.4 Ring-Type Joint Gaskets

The majority of metallic gaskets are ring-type joint (RTJ) gaskets, which are used in high pressure classes, such as class 600, which equals 100 bar pressure, as well as hazardous fluid services such as hydrocarbons. Gaskets manufactured by RTJ are precisely machined to fit within the grooves of mating flanges or other mechanical components. In the topside sector of the oil and gas industry, these gaskets are covered by ASME B16.20, and in the subsea sector, they are covered by API 6A. In Figure 2.36, an RTJ gasket is shown inside the groove of a flange.

2.6 BOLTING (BOLTS AND NUTS)

Fasteners (bolts and nuts) play a significant role in the reliability of flange connections. In the case of flanges, bolts are used that are called stud bolts, and nuts are hexagonal heavy nuts. Stud bolts are rods that are completely threaded and are used with nuts on each end to fasten them. An illustration of a stud bolt with two nuts is shown in Figure 2.37.

A standard for bolts and screws, ASME B18.2.1, contains general data and dimensions of bolts, such as the diameter and length of stud bolts. Furthermore, this standard specifies the dimensions and other information regarding square and hexagonal head bolts and screws. Stud bolts are threaded, headless fasteners mated with two nuts. A screw or bolt with heads, however, can be used to tighten two joints without the use of nuts. A standard for nuts, ASME B18.2.2, covers nut data, including hexagonal nuts for heavy loads. There is a slight difference in size and thickness between heavy hexagonal nuts and standard hexagonal nuts. The maximum thickness value of the heavy hexagonal nut is very close to the bolt size value. A standard for screw threads, ASME B1.1, governs the internal threads of nuts and the external threads of bolts. There are two abbreviations for unified form threads: UNC (unified coarse) and UNF (unified fine). Oil and gas industries are more likely to use UNC threads. When compared to UNF, UNC is able to handle a greater amount of load and allows for faster assembly and disassembly. According to the author's experience, UNC threads are commonly used for bolts with a diameter of 1" or less. However, 8UN threads, which have 8 threads per inch, are typically used for bolts with a diameter greater than 1". A washer may be used between the fasteners to improve the transfer of torque from the bolt to the fastener. After the flange joint has been prepared, the correct gasket has been selected and positioned between the flanges, and the bolts and nuts have been inspected and assembled to the flange joint, bolt tensioning should be applied in order to tighten

FIGURE 2.37 Stud bolts with nuts at each end.

the bolts. When tightening bolts, torque or tensioning force is applied. A torque is defined as the amount of twisting force required to spin a nut up along a bolt thread. It is important to note that there are two types of bolt-tightening tools. One is a torque tool, and the other is a bolt tensioning tool. For bolts with diameters of 1" and below, a torque tool is typically used, whereas a tensioning tool is used for bolts with diameters of 1 1/8" and above. There are two types of torque tools: manual and hydraulic. Hydraulic tensioning tools are typically used for the tensioning of bolts. Torque applied to bolts varies according to their size, their material, and the amount of lubrication applied to the bolts and nuts.

2.7 MECHANICAL JOINTS (HUBS AND CLAMPS)

There are many advantages to using mechanical joints instead of standard ASME flanges, including the ability to save weight and space. Mechanical joints are commonly used for high pressure class piping systems such as CL1500 and CL2500 in 3" size and larger. Mechanical joints consist of two hubs, two clamps, four stud bolts, eight nuts, and a seal ring (see Figure 2.38). The bolts, nuts, and seal rings used for mechanical joints do not comply with the typical ASME flange standards. Mechanical joints are available in three brands: G-Lok, Gray-Lok, and Tech-Lok.

There are many advantages to using mechanical joints in addition to saving weight and space. The following are other advantages of mechanical joints over ASME flanges:

1. Through a tight joint, high joint efficiency is achieved;
2. It is important to note that mechanical joints are resistant to a wide range of piping loads, including tension, compression, bending, and thermal shock. The manufacturer of the mechanical joints should provide a table indicating the joint's load capacity.
3. Unlike flanges, there is no internal fluid contamination.
4. Metal sealing is provided by mechanical joints with a soft cover, which provides tight sealing and is reusable, unlike gaskets.

FIGURE 2.38 A mechanical joint.

2.8 COMPACT FLANGES

Compact flanges are designed according to the Norsok L-005 or ISO 27509 standards and are typically larger and heavier than mechanical joints of the same size and pressure class but more compact and lighter than standard ASME flanges. As well as saving weight and space, compact flanges also have several advantages over ASME standard flanges, including better sealing, easier and faster installation and disassembly, excellent load capacity, and a reusable metal seal ring. Compact flanges have a much lower separation load than ASME flanges, which facilitates faster and simpler disassembly. Compact flanges are exceptionally effective at preventing leakage of internal fluid to the atmosphere because they provide double sealing against internal pressure, referred to as hill and wedge sealing. In addition, a compact flange prevents dirt or corrosive compounds from entering the interior of the flange through a "wedge" from the external environment. In Figure 2.39, a mechanical joint can be seen on the left, a compact flange in the middle, and a standard ASME flange on the right. There are certain disadvantages and limitations associated with compact flanges. Due to the sensitive nature of the flange face, it is not recommended to use a compact flange with rotary equipment. Additionally, compact flanges should not be used in applications that require frequent opening and closing.

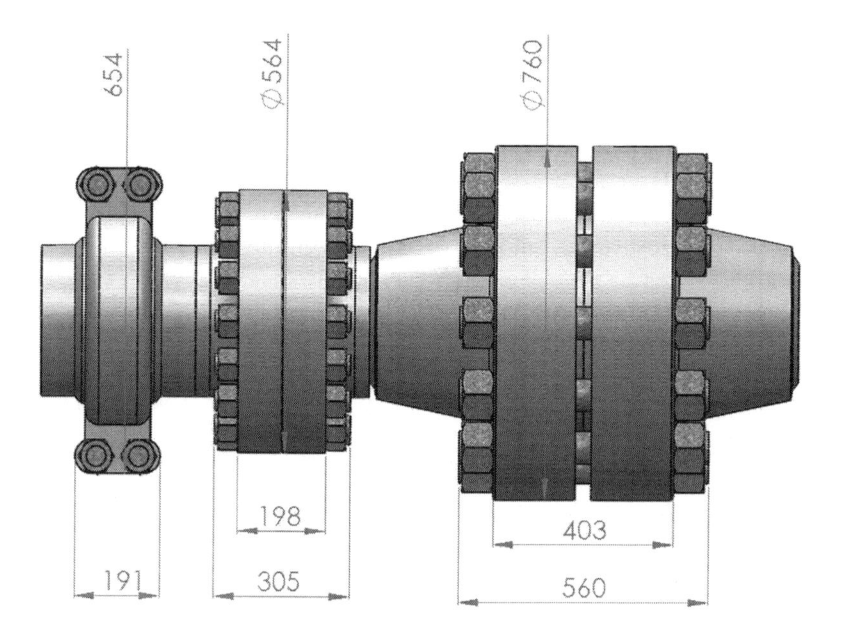

FIGURE 2.39 Size comparison between mechanical joint (left), compact flange (middle), and standard flange (right).

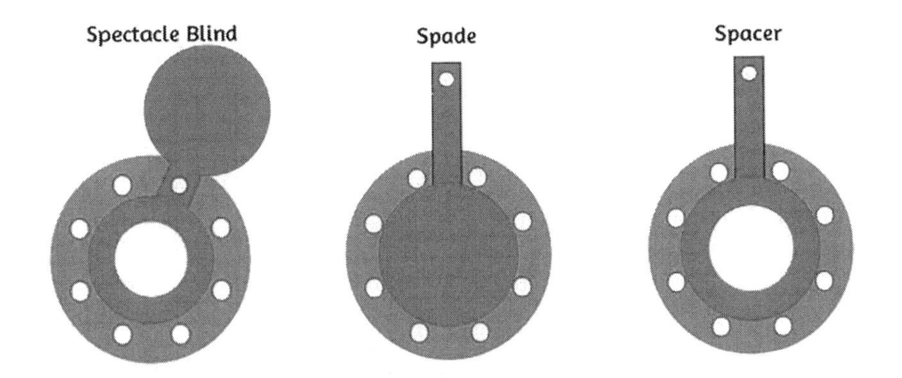

FIGURE 2.40 Spectacle blind, spade, and spacer.

2.9 SPADE AND SPACER/SPECTACLE BLINDS

In general, spectacle blinds are used to separate piping systems from each other for a variety of purposes, such as testing and maintaining piping lines. There are two plates with a specific thickness that form a spectacle blind. Of the two plates, one is solid, and the other has a ring-like hole in the center. The inside diameter of the ring is equal to the inside diameter of the flange connected to it. A spectacle blind is mounted between two flanges. Alternatively, a spade and spacer (Figure 2.40) can be used instead of a heavy spectacle blind.

2.10 PIPING SPECIAL ITEMS

2.10.1 Strainers

Strainers consist of a perforated plate and mesh element that are used to remove debris from fluids. It is recommended to place strainers upstream from pumps and compressors in order to remove particles such as sand from the fluid in order to prevent damage to the internal components of the equipment. Various types of strainers exist, such as Y-type, basket, tee (bath), and conical strainers (see Figure 2.41). Permanent strainers include tee strainers, Y strainers, and basket strainers. Conversely, conical strainers are classified as temporary strainers since they are temporarily installed upstream of sensitive equipment during startup and flushing. The purpose of temporary strainers is to protect equipment from construction debris that has been left in the pipe. The conical strainer (see Figure 2.42) is installed in the spool pipe between two standard flanges of the American Society of Mechanical Engineering. During commissioning, low-cost conical strainers are used temporarily. As opposed to other types of strainers, they do not have a pressure-retaining housing, and they are normally installed between flanges on a pipe spool. The filters consist of a cone-shaped perforated

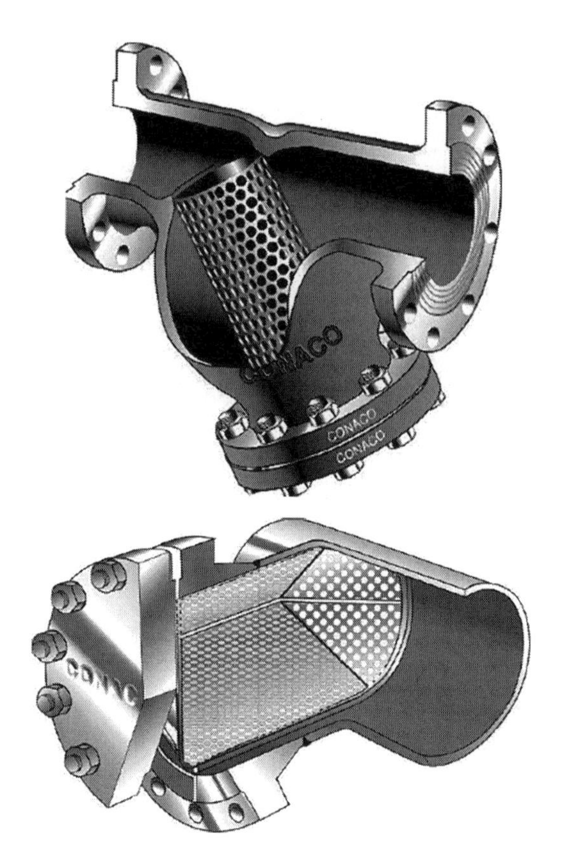

FIGURE 2.41 Y-type and tee strainers.

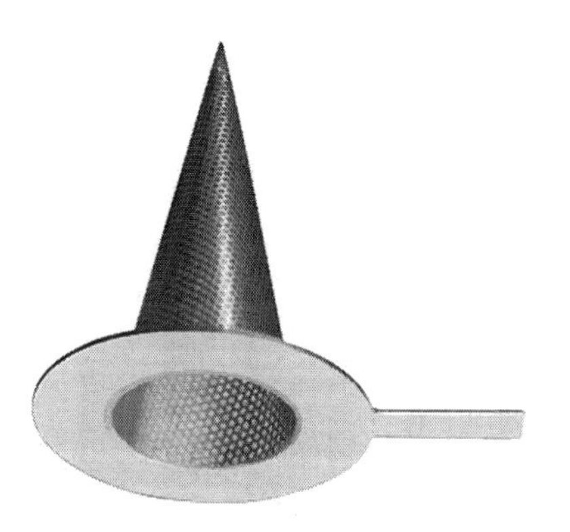

FIGURE 2.42 A conical strainer.

plate, sometimes in combination with a welded mesh to achieve the desired level of filtration. It is also possible to use a duplex strainer, which consists of two basket strainers connected in series. While this strainer has excellent filtration characteristics and continuous filtration during the basket change, it is more expensive and requires more space than other types of strainers.

The tee strainer (also known as the bathtub strainer) has a high pressure loss due to the fact that all debris is collected directly in the path of the process stream. With conical strainers, high pressure loss also occurs because all debris accumulates in the process stream, and any build-up that increases the pressure drop dramatically reduces the filtration effectiveness. However, Y-type and basket strainers have a very effective filtration area with a lower pressure drop and a higher flow open area because debris is collected away from the process stream. Compared to Y-type strainers of the same size and pressure class, basket strainers have a greater flow open area and a lower pressure drop. Nevertheless, tee-type strainers have the advantage of being less expensive than basket-type strainers. Additionally, both conical and tee strainers are suitable for temporary use during commissioning.

The size of a strainer is not necessarily the same as the size of the pipe. There should be consideration given to the amount of dirt and the strainer pressure drop. A strainer that is undersized requires frequent cleaning and can result in pressure drop issues in the system. It is important to consider the flow open area ratio, which indicates the ratio between the free flow area of the clean basket (screen) and the cross-sectional area of the pipe. For a clean screen strainer, the flow open area ratio should be at least 3 and at most 4.

2.10.2 Bends

The bends discussed in this chapter are considered special items of piping.

2.10.3 Manifolds

In the oil and gas industry, manifolds are commonly used to distribute process fluids such as oil, gas, and water. There are two types of manifolds: those that merge multiple junctions into a single channel and those that divide one flow line into multiple outputs. As an example, a production manifold is located before the separator, and it collects oil from different flow lines coming from wellheads. The produced fluid is transferred into a single channel, consisting of three phases of oil, gas, and water, for further processing in a separator. The fabrication of manifolds can be accomplished by welding wrought tees together or welding olets onto pipe headers.

2.10.4 Steam Traps

As the name implies, a steam trap is a device that discharges condensates and non-condensable gases without consuming or losing any live steam. A steam trap is nothing more than an automatic valve. They can be opened, closed, or modulated automatically. The operation of a steam trap is determined by the difference between the properties of steam and condensate. Because liquid condensate has a much higher density than gaseous steam, it will tend to accumulate at the lowest point in the steam system. Using this special item on a heating medium such as steam is beneficial in three ways: automatic discharge of condensate, prevention of steam leakage, and the ability to discharge non-condensate gases such as air. Therefore, steam traps are an integral part of any steam system. They are the key to good steam and condensate management, retaining steam within the process to maximize heat utilization and releasing condensate and incondensable gases at the most appropriate time. There is no limit to the pressure at which steam traps can operate, ranging from vacuum to more than 100 bar. Several types are available to meet these varied conditions, each with its own advantages and disadvantages. It has been demonstrated that steam traps work most efficiently when their characteristics are matched with those of the application. To perform a given function under given circumstances, the correct trap must be selected. Initially, these conditions may not appear obvious. It is possible that they may involve variations in operating pressure, heat load, or condensate pressure. It is possible for steam traps to be subjected to extreme temperatures. They may also need to withstand pressure and corrosion. Steam traps can be classified into three types: *mechanical traps*, *thermostatic traps*, and *thermodynamic traps*.

2.10.4.1 Mechanical Traps

The mechanical steam trap is designed to open for fluids that are more dense and close for fluids that are less dense. *Float traps* and *inverted bucket* traps are the two basic types of mechanical steam traps that operate on the density principle. There is a direct correlation between the level of condensate in the trap and the position of the float in float traps. By adjusting the valve to compensate for the flow of the condensate, the float responds to the flow of the condensate. Floating-ball traps (Figure 2.43) are similarly simple mechanical devices. As a

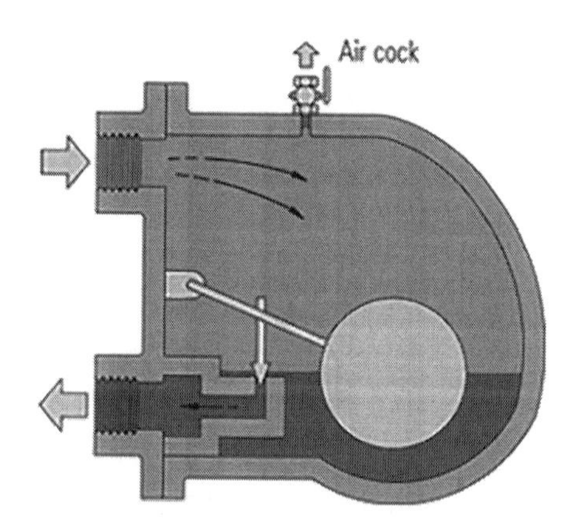

FIGURE 2.43 A floating-ball steam trap with lever.

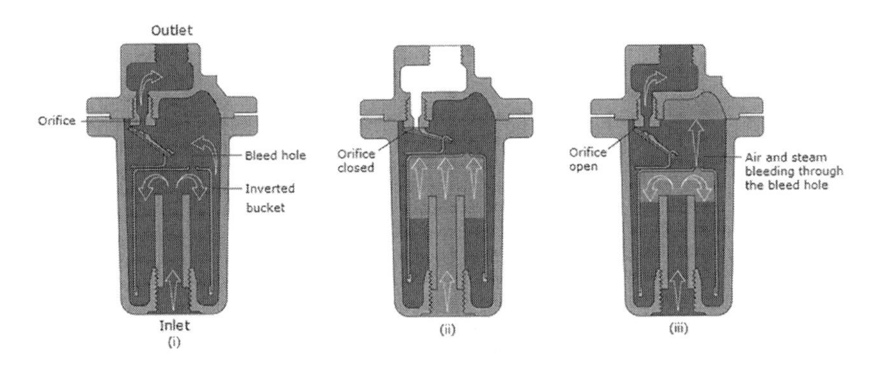

FIGURE 2.44 An inverted bucket steam trap.

result of the weight of the ball acting through a lever, the valve remains closed when there is no condensate present. In the event of condensate entering the trap, the ball will rise (float), and the valve will open as a result. After the condensate has been discharged, the ball returns to its original position and closes the valve. There are two basic types of float traps: lever floats and free floats. As part of lever float designs, a float is attached to a lever that controls the valve. Condensate entering the trap causes the float to become buoyant, causing the lever to move and the trap valve to open. In the free float type, the float is not attached to a lever, and the float itself serves as a valve. Floats are able to rise independently from the orifice, allowing condensate to be drained without obstruction. With inverted bucket steam traps, the bucket inside the trap is attached to a lever that opens and closes the valve in response to the bucket's motion. The steam generated by the flow of steam or air into the underside of the inverted bucket, combined with the condensate surrounding the bucket on the outside, causes it to be buoyant and rise (see Figure 2.44). By placing the bucket in this position, the trap valve will be closed. An important difference between float traps and inverted bucket traps is the type of condensate drainage they provide; float traps provide continuous drainage, while inverted bucket traps provide intermittent drainage.

2.10.4.2 Thermostatic Steam Traps

A thermostatic trap operates according to the temperature of the steam passing through the trap. There are three major types of thermostatic traps, *liquid expansion traps* and *bimetallic* and *balanced pressure thermostatic* traps.

A liquid expansion steam trap is a thermodynamic trap that uses steam temperature to expand liquid inside a bellow in order to open or close the internal valve. One of the simplest thermostatic traps is the liquid expansion steam trap. The trap consists of an oil-filled element that expands when heated due to steam entering the trap during the heating process. By doing so, the internal valve will close against the seat. Figure 2.46 shows a liquid expansion steam trap.

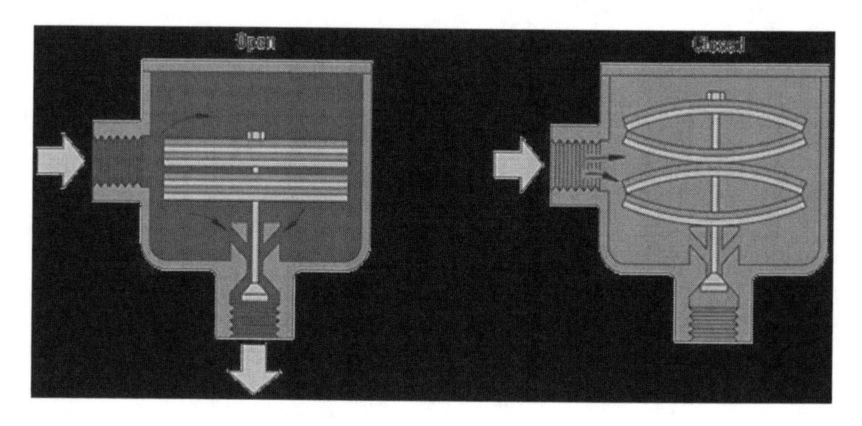

FIGURE 2.45 A bimetallic steam trap.

2.10.4.2.1 Bimetallic Steam Traps

In order to construct a bimetallic steam trap, two strips of dissimilar metals are welded together into one piece. Heat causes the element to deflect. The steam trap that we are describing is a temperature-operated trap, also known as a thermostatic steam trap. It takes advantage of the fact that steam has a higher temperature than condensate. In order to close this steam trap, a higher temperature of steam is used. There is no doubt that this is the most robust design of steam trap among all steam traps. Bimetallic strips of dissimilar metals are stacked in pairs in this trap. The left side of Figure 2.45 shows a steam trap in an open state. As long as the steam trap contains condensate, the bimetallic strips are in the flat position, and the valve is open. The condensate drains freely from the system. As steam enters the trap, its higher temperature causes the bimetallic strips to bend and close the valve. As a result, no steam is expelled from the system.

2.10.4.2.2 Balanced Pressure Thermostatic Steam Traps

Steam traps with balanced pressure are temperature-operated traps, which means that when steam enters the trap, a higher temperature than the condensate is used to close the trap. A bellows diaphragm is used in balanced pressure thermostatic traps. Upon application of pressure and temperature, the diaphragm opens to discharge the air and condensate and closes to prevent the passage of steam.

2.10.4.3 Thermodynamic Steam Traps

A thermodynamic steam trap is a simple and compact device that consists of a single moving component (disc), which opens in the absence of steam and closes when the presence of steam is detected.

2.10.5 Expansion Joints

The purpose of expansion joints is to absorb vibrations and shocks in a piping system. In addition to reducing the noise, they are also capable of compensating for

FIGURE 2.46 An expansion joint.

misalignments. It is also possible to design special expansion joints for applications that require thermal expansion. An expansion joint is shown in Figure 2.46. Metal, rubber, or braided expansion joints are the most common types of expansion joints. In most cases, metal expansion joints are used in applications where thermal expansion is a concern. As the temperature of the pipe rises, the metal expansion joint compresses in order to compensate for the movement and relieve the pipe of stress. A variety of materials are available for metallic bellows, including stainless steel and nickel alloys. Flexible rubber expansion joints are made of elastomers with metallic reinforcement for providing stress relief in piping systems during thermal changes. Rubber expansion joints are extremely effective at absorbing vibrations and shocks. As a result of this type of expansion joint, noise can also be reduced.

2.10.6 Flexible Hoses and Couplings

In fluid transportation or transfer, a flexible hose is a type of piping special item used to connect two distant points. Hoses are used in oil and gas applications when there is a great deal of relative movement. Through flexible hoses, fluids and fluidized solids can be easily transferred from one location to another. The most commonly known of these are hosepipes. When flexibility is needed to handle vibration, thermal expansion, or construction issues such as ease of routing, flexible hose is preferred over rigid piping. It is not uncommon for oil companies to use flexible hoses to transfer crude oil and other products between ships or barges and shoreside facilities. A flexible hose is connected to a line through a coupling or flange. Flange connections are more robust and can handle higher pressures and more aggressive fluids, such as hydrocarbons and chemicals. Additionally, hoses with flange connections should be selected for connection to equipment such as pumps and compressors. Flexible hoses and couplings are shown in Figure 2.47.

It is possible to manufacture hoses from rubber, composite materials, or metallic materials. A metallic hose, for example, could be constructed from stainless steel. The flexibility of non-metal hoses, such as those made of Teflon, is greater than that

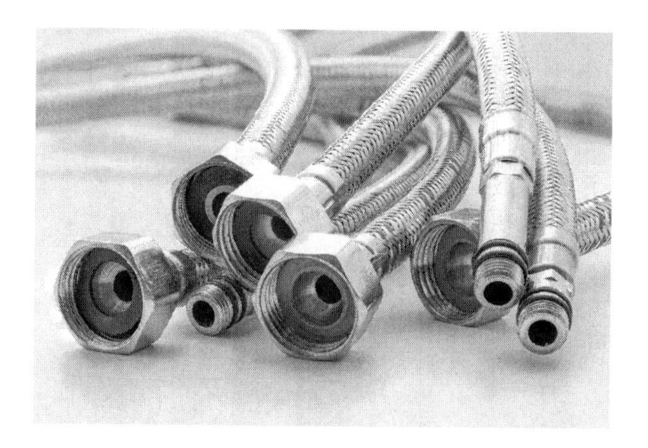

FIGURE 2.47 Flexible hoses and couplings.

(Courtesy: Shutterstock)

of metallic hoses. The operating temperature is another factor that influences the selection of hose materials. The use of non-metallic hoses is not recommended for high-temperature applications. Metallic hoses are the only material suitable for use in extreme cold or high temperatures, regardless of the temperature of the media entering the hose or the surrounding atmosphere. It is essential that rubber hoses be flexible and capable of being bent without being overstressed. The flexibility of a hose made of a non-metallic material, such as Teflon, is greater than that of a hose made of metal. A hose used in vacuum services must also be stiff and strong enough to resist collapse. Due to this strength, metallic hoses are less likely to collapse during vacuuming operations. The strength of metallic hoses in abrasive services containing particles is greater than that of non-metallic hoses. In the event of a fire, soft materials would not be fire safe and would melt. However, some metallic hoses are capable of withstanding temperatures up to 700°C. Fluid velocity is another factor that needs to be considered when choosing hose materials. In order to minimize the risk of erosion, metallic hoses should be used with high fluid velocity.

2.10.7 Sample and Injection Quills

Both sample quills and injection quills have essentially the same design; however, as their names suggest, one is used to inject chemicals into a pipeline, tank, or process, and the other is used to extract or sample samples from a process system for analytical purposes. A sampling quill is used for safely removing fluid samples from a process line for analysis or laboratory testing. It is mounted into the pipe through a fitting in a manner that allows the quill/probe to be inserted into the pipe and extend down to the center of the process line. A chemical injection quill is a mechanical device used in many industries to inject chemicals. This device, as illustrated in Figure 2.48, is used to interface chemical feed lines with process piping.

FIGURE 2.48 An injection quill.

2.10.8 Backflow Preventors

There is a constant threat to drinking water quality from backflow contamination incidents, whether they are nuisances, non-health hazards, or serious public health emergencies. In a backflow preventer, two check valves are connected in series with a release or bleed valve in the middle (see Figure 2.49). The installation provides double isolation between polluted downstream water and potable water upstream to prevent the spread of disease. According to the level of protection required, different types of backflow preventers are selected. There may

FIGURE 2.49 A backflow preventor.

(Courtesy: Shutterstock)

FIGURE 2.50 Bird screens.

be a strainer upstream and two isolation valves downstream and upstream of the strainer in a backflow preventer.

2.10.9 Bird Screens

The purpose of bird screens (or insect screens) is to prevent birds or insects from being trapped inside an open-ended pipe (such as an exhaust pipe), where they can cause a blockage. It is common for birds such as pigeons and other roosting birds to nest inside exposed piping, and their droppings can carry diseases and even pose health hazards. In addition to causing damage to sensitive equipment such as valves and gauges, bird droppings have also been known to create safety hazards on ladders, steps, and catwalks. It is best to incorporate bird screens into the refinery, chemical, power, LNG, or green energy plant design and to install them during construction. Figure 2.50 shows various bird screens used in oil and gas plants.

2.10.10 Vacuum Breaker and Air Release Valves

Vacuum breaker and air release valves (see Figure 2.51) have two functions: vacuum breaker and air release. In vacuum breaker mode, the orifice will be open to allow air to enter, and the liquid will drain to prevent damaging the pipe. The air release function is provided by this type of valve. It is capable of supplying and removing air from the pipeline as necessary. It is capable of venting the pipeline. This device is capable of removing accumulated air at high points during pressurized operation of the system. In a vacuum breaker valve, large quantities of air are automatically removed during pipeline filling, and large quantities of air can also be received automatically when the internal pressure in the pipeline falls below atmospheric pressure. At start-up, the air release function of the valve automatically vents a large volume of air out of the tank or piping system.

2.10.11 Excess Flow Valves

Excess flow valves are mechanical safety devices installed on natural gas service lines to prevent the flow of natural gas from continuing if there is a significant

FIGURE 2.51 Vacuum breaker and air release valves.

break, puncture, or severance in the line. It is likely that more flow will pass through these valves if a pipe break occurs after these valves. Spring-loaded flow shut-off valves are designed to close only when the flow through the valve exceeds a predetermined closing flow rate. Depending on the conditions, each valve has a specific closing flow rate.

2.10.12 Tube Connectors

Fittings and connectors for tubes may be manufactured by Parker or Swagelok. Installation, assembly, and disassembly are straightforward. It is possible to thread the tube connection from one side of the valve to the body of the valve and connect the tube to the valve through two sleeves. Figure 2.52 shows a tube connection manufactured by Swagelok. There is a threaded male end at the bottom that can be attached to a female fitting or body valve.

There are four components to the tube connection: the body, the nut, the front ferrule, and the back ferrule. A body determines the shape and type of connector on the end, which may be female or male. It is the nut that creates the force between the tubing and the ferrules. A front ferrule creates a seal around the outside diameter of the tubing. Tubes are held in place by the back ferrule (smaller).

2.10.13 Special Tees (Barred Tees)

Generally referred to as pigging or scraping, pipeline inspection gauges or gadgets are used to perform various maintenance operations on pipelines. A barred tee, or piggable tee, is a special pipe fitting used in piping injected gadget (PIG) launchers and receivers on a pipeline so that the PIG can pass the tee without

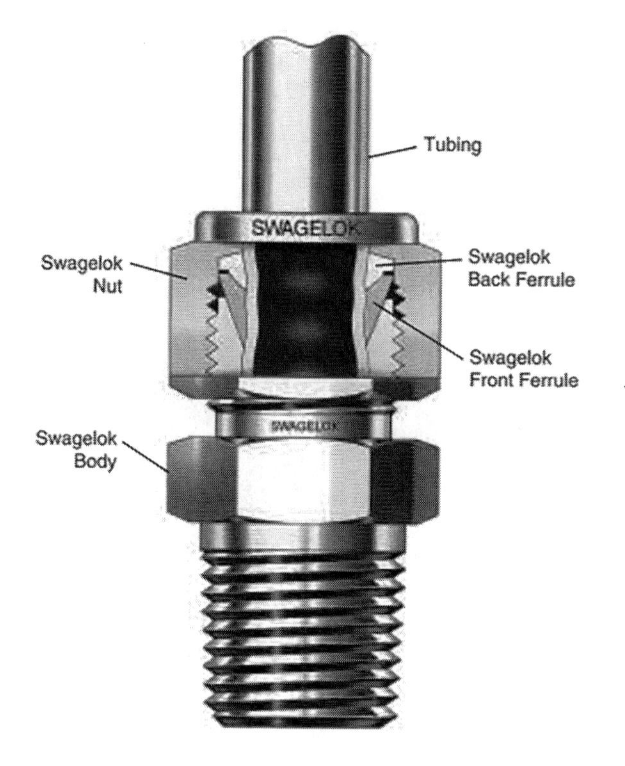

FIGURE 2.52 Tube fitting.

moving to the branch connection. Barred tees are the same shape as regular tees; however, there are bars welded to the branches to prevent the PIG from running into them. A barred tee for a pipeline is shown in Figure 2.53. The bar material welded to the internal diameter of the branch connection should be the same as the material used for the tee. There are bars placed (welded) on the internal sides of the tee. It is important that the bars on the branch be small enough to avoid creating any flow restrictions on the branch connection. However, they should not be so small and weak that they may break due to the flow rate through the branch connection to the tee. Barred tees are made based on ASME B16.9, "Factory Made Wrought Butt-Welding Fittings," or MSS SP 75, "Specification for High-Strength, Wrought, Butt-Welding Fittings."

2.10.14 Insulation Kits

An insulation kit (see Figure 2.54) consists of gaskets, sleeves, and washers for insulating two flanges made of dissimilar materials in order to prevent galvanic corrosion. The galvanic corrosion process (also known as bimetallic corrosion or dissimilar metal corrosion) occurs when one metal corrodes preferentially when

FIGURE 2.53 A barred tee.

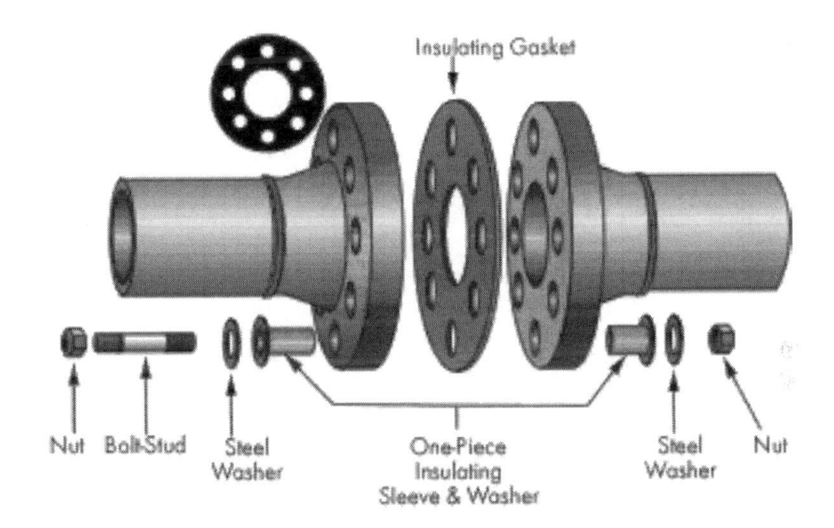

FIGURE 2.54 An insulation kit between two flanges in dissimilar materials.

it is electrically connected to another in the presence of an electrolyte. By virtue of the movement of its ions, an electrolyte is a medium containing ions that is electrically conductive. Gaskets are used between flanges in two different types. A type E gasket, such as the one shown in the figure, extends beyond the outside diameter of the flange and has holes inside for the passage of bolts. In addition to phenolic or rubber-faced phenolic gaskets, other non-metal materials, including PTFE, Viton, and nitrile may also be used. Type F gaskets extend up to the bolts on the flange. It is important to note that the outside diameter of the type F gasket extends beyond the internal diameter of the bolt hole circle.

2.10.15 Flame Arrestors

There are several names for flame arresters, including deflagration arresters, flame traps, and flame arrestors. As the name implies, a flame arrester is a device or construction that permits the passage of gases or gaseous mixtures but prevents the passage of flames. It prevents flames from being transmitted through a flammable gas/air mixture by extinguishing them on the high surface area provided by a series of small passages through which they pass. In the protected area, the emerging gases are sufficiently cooled to prevent ignition. Sir Humphry Davy, a celebrated chemist and professor at the Royal Institution in England, discovered the principle of flame arresters in 1815. Numerous types of flame arresters have been used in many industries since Sir Humphry's time. They all work on the same principle: preventing the flame passage and removing heat from the flame as it passes through narrow passages with metal walls or other heat-conductive materials. A flame arrestor may be installed at the end or opening of a pipe or inside a piping system (inline). An end-of-line flame arrestor is shown in Figure 2.55.

2.10.16 Special Flanges and Gaskets

As the name implies, special flanges are those not covered by flange standards such as ASME B16.5, ASME B16.47, and MSS SP 44. In pressure classes from 150 to 1500, ASME B16.5 covers flanges up to and including 24". In pressure classes 2500, ASME B16.5 covers flanges up to and including 12". The ASME B16.47 standard covers flanges of larger sizes ranging from 26" to 60" in diameter. According to ASME B16.47, there are two different series of A and B that

FIGURE 2.55 An end-of-line flame arrestor.

(Courtesy: Shutterstock)

are not interconnected. Per ASME B16.47, Series A flanges have larger bolt holes and fewer bolt holes than Series B flanges. Series B has smaller bolts that provide better integrity than Series A. MSS SP 44 flanges can be compared with ASME B16.47 Series A. ASME B16.47 Series B does not cover sizes larger than 36" for classes 400 and higher. It is therefore necessary to manufacture a flange in 40" and class 1500 as a special item. The internal bore of a flange may be selected according to the flow capacity required by the process department. Finite-element analysis of flanges for high pressure classes should be performed in order to ensure that they are strong enough to withstand the loads. The ASME boiler and pressure vessel section VIII Appendix II can be used to calculate loads for special flanges. Earlier in this chapter, we discussed reducing and expander flanges as examples of special flanges. There may be a need for special flanges in connection with the equipment. The internal diameter (ID) of the equipment flange can be larger than the ID of the connected flange. As a result, the bore of the connected flange should be tapered to achieve a special bore to match the nozzle of the equipment. As mentioned earlier, reducing flanges (e.g. 8" × 6") are 8" flanges with a 6" bore, and the spiral wound gasket should be configured differently between two reducing flanges. An inner ring must be compatible with a 6" bore pipe and a 6" spiral wound gasket, and an outer ring should extend to an internal diameter of an 8" flange bolt hole. It may not be necessary to use a special gasket in some cases.

2.10.17 Transition Pieces

When there is a high thickness differential between two piping components, it is not possible to weld them together directly. In such a case, a transition piece is welded between two components to serve as a type of joint. The thickness of one end of the transition piece matches that of the thicker component, and the thickness of the other end matches that of the thinner component.

QUESTIONS AND ANSWERS

1. There is a high level of hydrogen sulfide in the fluid in the piping service, and the pipe is exposed to severe cycling conditions. Process department personnel are concerned about fluid turbulence, pressure drop, and erosion in the given piping system. The concern may arise from the use of a pipe routing change after the pressure relief valves or after the compressors where pressure drop is a concern. It is important to be aware that there are no restrictions on the layout of the pipe. If these conditions are met and 90° rotation is required for such piping, which piping component is the most appropriate?
   A. Short radius elbow
   B. Bend
   C. Long radius elbow
   D. Miter bend

Answer) Considering that the fluid service is very corrosive and the pipe is exposed to severe cycling stresses or loads, a miter bend is not an appropriate choice. Furthermore, the miter bend would increase fluid turbulence and pressure drop in comparison to the other options. Therefore, option D is incorrect. Generally, short radius elbows are not recommended because they create a high pressure drop, so option A is also not recommended. It is not advisable to use an elbow with a long radius due to concerns regarding flow turbulence and pressure drop raised by the process department. Another option is to use a bend with a relatively long radius, such as 3D or 5D, which provides smooth flow with relatively low turbulence. A bend requires more space than the other options, but since the piping routing has no layout limitations, a bend is recommended for such an application; therefore, option B should be selected.

2. Which piping component is suitable for blinding the end of the pipe?
   A. Reducer
   B. Swage
   C. Tee
   D. Cap

   Answer) Option D is the correct answer.

3. Which sentence is completely correct and complete about olets?
   A. The welding of an olet to piping is more complex than the welding of a tee.
   B. A sockolet is used to connect a 6" header to a ½" branch, and its pressure class is 6000, forged A105, which is made according to MSS SP 97.
   C. The olet is always connected to the branch by a fillet weld.
   D. A threadolet is connected to the header through a thread connection.

   Answer) Option A is incorrect since welding the tee to the piping is more complex than welding an olet. The correct answer is option B. Due to the fact that the olet is connected to the header through a filet weld, option C is incorrect. In addition, option D is incorrect because the threadolet is attached to the header via a fillet weld and to the branch via a thread connection.

4. Which standard is not related to the allocated component?
   A. A swage is made based on the MSS SP 95 standard.
   B. Olets are covered according to the ASME B16.9 and ASME B16.11 standards.
   C. Reducers are covered by either the ASME B16.9 or ASME B16.11 standard.
   D. Induction bends are made according to the ASME B16.49 standard.

   Answer) Option B is not completely correct, as olets are made according to the MSS SP 97 standard.

5. Which component cannot reduce the size of the pipe?
   A. Concentric reducer
   B. Eccentric reducer
   C. Equal tee
   D. Swage

   Answer) Option C is the correct answer.

6. Which sentences are not correct statements about gaskets?
   A. Spiral wound gaskets should be handled with extra care to prevent scratches and damage to the graphite filler.
   B. Asbestos materials are allowed for flat gaskets and for the filler of spiral wound gaskets.
   C. RTJ gaskets are used for high pressure classes, such as CL600 and higher.
   D. The maximum hardness of RTJ gaskets is not important.

   Answer) It is correct to choose option A, because filler gaskets are at a high risk of being scratched and removed during handling. Option B is incorrect, since asbestos materials may cause serious health problems, such as lung cancer, so their use is prohibited. Option C is correct; RTJ gaskets are used for high pressure class flanges such as CL600, and sometimes CL900 and above. Option D is incorrect, as metallic RTJ gaskets should be softer than the flanges in order to deform properly and provide sealing. According to ASME B16.20, RTJ flanges in different materials have maximum hardness values.

7. Find the incorrect statement regarding barred tees.
   A. Barred tees are considered special piping items.
   B. For pigging operations in the main pipeline header, a barred tee should provide an unobstructed profile.
   C. Ideally, the bars should have a higher material quality than the body and be welded in a manner that is compatible with the body of the tee.
   D. To prevent the entry of PIG spheres into the branch pipes, guide bars are required.

   Answer) Option C is not completely correct, since the material of the bar and the body of the barred tee should be the same.

8. Which piping special item is used to prevent corrosion?
   A. Flame arrestor
   B. Insulation kit
   C. Steam trap
   D. Strainer

   Answer) Option B is the correct answer.

9. In a specific size and pressure class, which component is more compact?
   A. Compact flange
   B. ASME B16.5 flange

C. Hub and clamp
D. ASME B16.47 flange

Answer) Option C is the correct answer.

10. A floating ball and an inverted bucket belong to what category of special items?
    A. Flame arrestor
    B. Sample quill
    C. Steam trap
    D. Strainer

    Answer) Option C is the correct answer.

FURTHER READING

1. American Petroleum Institute (API) 5L. (2018). *Line pipe*. 46th ed. Washington, DC: API.
2. American Society of Mechanical Engineers (ASME) Section VIII Div. 01. (2012). *Rules for construction of pressure vessels: Boiler and pressure vessel code*. New York, NY: ASME.
3. American Society of Mechanical Engineers (ASME) Section VIII Div. 02. (2012). *Design and fabrication of pressure vessels: Boiler and pressure vessel code*. New York, NY: ASME.
4. American Society of Mechanical Engineers (ASME) B 31.3. (2010). *Process piping*. New York, NY: ASME.
5. American Society of Mechanical Engineers (ASME) B1.1. (2019). *Unified inch screw threads (UN, UNR and UNJ thread forms)*. New York, NY: ASME.
6. American Society of Mechanical Engineers (ASME) B16.5. (2017). *Pipe flanges and flanged fittings: NPS ½" through NPS 24 metric/inch standard*. New York, NY: ASME.
7. American Society of Mechanical Engineers (ASME) B16.9. (2018). *Factory-made wrought buttwelding fittings*. New York, NY: ASME.
8. American Society of Mechanical Engineers (ASME) B16.11. (2005). *Forged fittings, socket-welding and threaded*. New York, NY: ASME.
9. American Society of Mechanical Engineers (ASME) B 18.2.1. (2012). *Square, hex, heavy hex, and askew head bolts and hex, heavy hex, hex flange, lobed head, and lag screws (inch series)*. New York, NY: ASME.
10. American Society of Mechanical Engineers (ASME) B 18.2.2. (2015). *Nuts for general applications: Machine screw nuts, hex, square, hex flange and coupling nuts (inch series)* New York, NY: ASME.
11. American Society of Mechanical Engineers (ASME) B36.10M. (2015). *Welded and seamless wrought steel pipe*. New York, NY: ASME.
12. American Society of Mechanical Engineers (ASME) B36.19M. (2018). *Stainless steel pipe*. New York, NY: ASME.
13. American Society of Mechanical Engineers (ASME) B16.20. (2007). *Metallic gaskets for pipe flanges, ring joint, spiral wound, and jacketed*. New York, NY: ASME.
14. American Society of Mechanical Engineers (ASME) B16.21. (2016). *Non-metallic flat gaskets for pipe flanges*. New York, NY: ASME.

15. American Society of Mechanical Engineers (ASME) B16.25. (2017). *Buttwelding ends*. New York, NY: ASME.
16. American Society of Mechanical Engineers (ASME) B16.36. (2020). *Orifice flanges*. New York, NY: ASME.
17. American Society of Mechanical Engineers (ASME) B16.47. (2017). *Large diameter steel flanges: NPS 26" through NPS 60" metric/inch standard*. New York, NY: ASME.
18. American Society of Mechanical Engineers (ASME) B16.49. (2017). *Factory-made, wrought steel, buttwelding induction bends for transportation and distribution systems*. New York, NY: ASME.
19. American Society of Mechanical Engineers (ASME) B1.20.1. (2013). *Pipe threads, general purpose, inch*. New York, NY: ASME.
20. Jaszak, P., & Adamek, K. (2019). Design and analysis of the flange-bolted joint with the respect to required tightness and strength. *Open Engineering*, 9(1). https://doi.org/10.1515/eng-2019-0031
21. Manufacturers Standardization Society (MSS) SP 95. (2018). *Swaged nipples and bull plugs*. [online] available at: https://www.techstreet.com/standards/mss-sp-95-2018?product_id=2030897
22. Manufacturers Standardization Society (MSS) SP 97. (2019). *Integrally reinforced forged branch outlet fittings—socket welding, threaded, and buttweld ends*. [online] available at: https://www.techstreet.com/standards/mss-sp-97-2019?product_id=2042157
23. Marston, E. (2000). *Piping joints handbook*. BP Amoco Document No. D/UTG/054/00. [online] available at: https://www.academia.edu/8161263/Piping_Joint_Handbook
24. Mohinder, L. N. et al. (2000). *Piping handbook*. 7th ed. New York, NY: McGraw-Hill.
25. Nayyar, M. L. (2000). *Piping handbook*. 7th ed. New York, NY: McGraw Hill.
26. Norsok Standard L-005. (2006). *Compacted flanged connections*. 2nd ed. Lysaker, Norway: Standard Norge.
27. Smith, P. (2007). *The fundamentals of piping design. Chapter 2: Piping components* (pp. 49–113). Houston: Gulf Publishing.
28. Sotoodeh, K. (2020). Manifold technology in the offshore industry. *American Journal of Marine Science*, 8(1), 14–19. https://doi.org/10.12691/marine-8-1-3
29. Sotoodeh, K. (2021). *A practical guide to piping and valves for the oil and gas industry*. Houston: Gulf Professional Publishing.
30. Sotoodeh, K. (2019). Handling the pressure drop in strainers. *Journal of Marine System and Ocean Technology*, 14, 220–226, Springer. https://doi.org/10.1007/s40868-019-00063-2
31. Sotoodeh, K. (2022). *Piping engineering: Preventing fugitive emission in the oil and gas industry*. 1st ed. New York, NY: Wiley. ISBN: 978-1-119-85203-2
32. Roy, A. et al. (2012). *Pipe drafting and design*. 3rd ed. Oxford: Elsevier Science.
33. Woods, G. E., & Baguley, R. B. (2001). *CASTI guidebook ASME B31.3: Process piping*. 3rd ed. Alberta, Canada: Casti Publishing.

3 Valves

3.1 INTRODUCTION

A valve is an important component of a piping system and is used for many purposes, including stopping or starting the flow of fluid inside the pipe, also referred to as on/off applications; regulating or controlling the flow of fluid; preventing backflow; and performing safety functions. About 20% to 30% of the total cost of piping can be attributed to valves. It is imperative that process piping engineers be knowledgeable about all types of valves and the introduction of any new designs. A thorough understanding of how valves operate, are maintained, and are adjusted is equally important for the success of a process plant and its overall operation. All valves, with the exception of check valves and pressure safety valves, can be either manual or automated. A manual valve is operated by the operator moving a lever or handwheel on the valve to open and close it. Nevertheless, some valves are automatically opened and closed by means of actuators. It is possible to automate valves via actuators for a variety of reasons, such as ease of operation, speed of operation, or automatic operation. There are some valves located in hazardous areas where personnel are at risk. It is possible to operate the actuator pneumatically, hydraulically, or electrically. In the next chapter, you will find more information about actuators. The reliability and good performance of the valve are very important. It is common for industrial valves to fail for a variety of reasons, including poor valve selection, a lack of mechanical strength against pipeline loads or fluid pressure inside the piping, poor material selection, corrosion, inappropriate coating or coating application, sealing failures, or poor testing and inspections. In the oil and gas industry, industrial valves that fail to function are a major risk and costly phenomenon with severe negative consequences, including the loss of assets; the loss of production as a result of plant shutdown; the loss of human lives; and health, safety, and environment (HSE) issues such as pollution of the environment. In general, valve design involves many different considerations, including the selection of materials, calculations of wall thickness, and the design of pressure-controlling internal valve parts, as well as the design of pressure-containing valve components, including the body, bonnet, stem, bolts, and sealing. In the event of valve failure, pressure-containing parts cause external leakage from the valve, while pressure-controlling parts affect the passage of fluid through the valve.

3.2 ON/OFF VALVES

In the oil and gas industry, on/off valves are used to start and stop fluid flow. There are four main types of on/off valves: ball valves, gate valves, plug valves,

DOI: 10.1201/9781003465881-3

and butterfly valves. There are two main types of gate valves: through conduit gate (TCG) valves and wedge-type gate valves. Single isolation is provided by the four main types of valves. A modular valve is another type of valve that is used for double isolation. In the following section, we will provide a more detailed explanation of ball valves.

3.2.1 Ball Valves

One of the most widely used types of valves in the world is the ball valve. From heavy-duty industrial applications to household plumbing, they are suitable for a wide range of applications. A ball valve is a flow control device that utilizes a hollow, perforated, pivoting ball in order to control fluid flow. An open valve is one in which the hole through the middle of the ball aligns with the flow inlet, and a closed valve is one in which the handle pivots 90°, preventing the flow of fluid. A ball called a closure member moves 90° between open and closed positions. It is not feasible to use ball valves for fluid control, also known as throttling, which would require keeping the ball in an intermediate position between open and closed. The reason a ball valve cannot be used for this purpose is that it would be subject to excessive wear and damage. A ball valve in open position is shown in Figure 3.1; here, the hole inside the ball aligns with the flow direction, allowing fluid to pass through it. This half ring is referred to as the seat in the picture. It is shown between the valve shell or body and the ball. Valve stems are also important components, as they are connected to the valve ball from one side and to the valve operator (gear or actuator) from the other. During valve opening and closing, the stem transfers the load from the operator to the valve closure

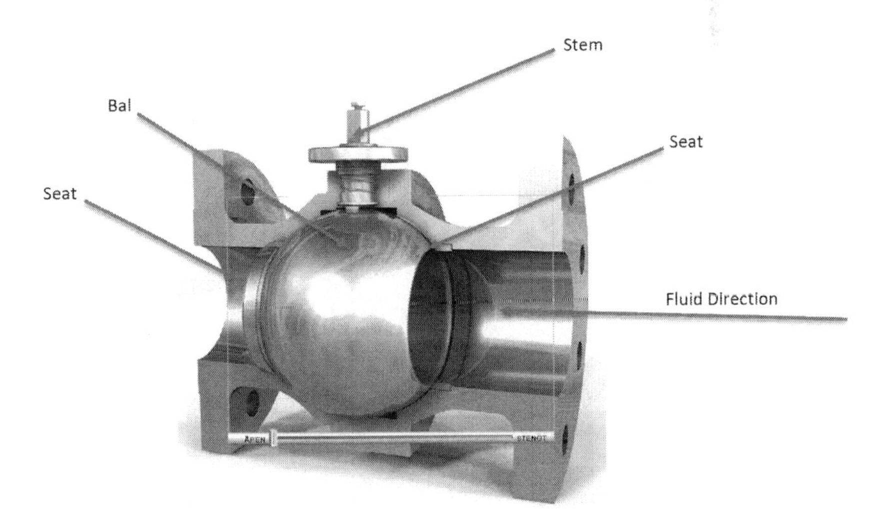

FIGURE 3.1 Ball valve in open position.

member. A ball valve is also known as a quarter-turn valve, as the ball and stem move only a quarter turn, or 90°, between open and closed states. As a result of the quarter-turn movement of the ball, a ball valve is capable of delivering fast operation and flexibility.

Valve bodies or shells, including ball valves, provide sealing between the fluid and its external environment by containing pressure. In the event of a failure of the body, internal fluid is released into the environment. Aside from the body, the stem and bolting (bolts and nuts) are also considered pressure-containing parts. In ball valves, bolts are used to connect the body pieces or the body and bonnet. Bonnets are valve components that are installed on the body to provide cover for the valve. In Figure 3.2, the ball hole of a ball valve is perpendicular to the flow direction when the valve is closed.

In addition to clean fluids, ball valves can also be used with dirty or particle-containing fluids. Ball valves used in clean fluids differ from those used in particle-containing fluids primarily because of the seat material and design; a metallic seat is used in dirty fluids, while a soft seat is used in clean fluids. Other reasons may justify the selection of a metal-seat ball valve, such as applications that are prone to high temperatures and significant pressure drops. In applications where dirty fluids are used, as well as in applications where there is a high level of pressure drop and high temperatures, a soft or non-metallic seat may be damaged. In other applications, soft-seat ball valves could be chosen first since they are less expensive than metal seats, provided that the mentioned conditions with regard to

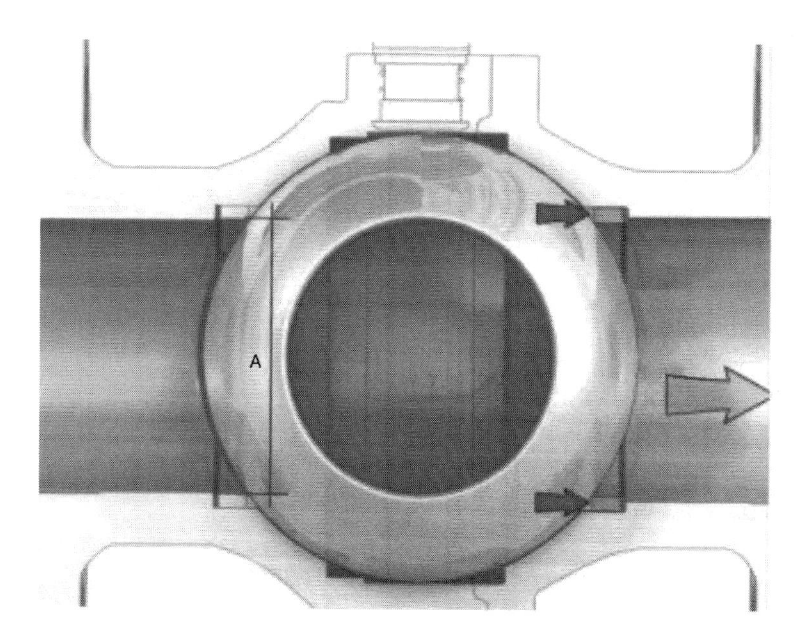

FIGURE 3.2 Ball valve in closed position.

(Source: photograph by author)

fluid cleanliness, high temperature, and pressure drop do not exist. In Figure 3.3, a metal-seat ball valve is shown in the half-open position.

Depending on their size, ball valves can be either full bore, reduced bore, or special bore. Full-bore ball valves have a straight flow path within them; the internal bore of the valve corresponds to both ends connected to piping. Alternatively, a full-bore valve may be referred to as a full port valve. A full-bore ball valve has the benefit of minimizing pressure drops and fluid capacity losses across the valve as well as less wear on the internal components of the valve. In a reduced-bore ball valve, the internals of the valve and its inside diameter are less than the diameter of the piping connected to it, for example, a reduced-bore ball valve with 3" flanges and a reduced bore of 2". Reduced-bore ball valves are identified and designated by two sizes separated by an ×. A valve has two sizes: one is the larger size at the point of connection to the piping, and one is the smaller size inside the valve (e.g. 3" × 2"). In the case of sizes 12" and smaller, the bore of the ball valve may be one size less than the size of the valve's end connection; for example, a 12" end connection with a bore size of 10" could be selected. Valve size can be written as 12" × 10". It is possible to reduce the bore by two sizes for sizes of 14" and larger. In the case of a 14" reduced bore ball valve connected to a 14" line, the valve's ends would be 14" and the bore reduced to 10", which is two sizes smaller than 14". This valve has dimensions of 14" × 10".

When compared with a full-bore ball valve, a reduced-bore ball valve has a smaller ball and smaller internals and requires less force or torque to open and close. A reduced-bore ball valve is typically the first choice when selecting a ball valve in sizes above 2", since it is more economical and less expensive than a full-bore ball valve. Valve and process engineers select valves based on a variety of parameters, such as size, pressure, temperature, fluid type, and application. Reducing the bore of a ball valve is generally more economical than increasing

FIGURE 3.3 Metal-seat ball valve.

(Courtesy: Shutterstock)

the bore, but there are some conditions under which a reduced-bore ball valve cannot be used. Pressure drop and wear inside the valve are the first two conditions. A reduced-bore ball valve should not be used if the pressure drop caused by it is high enough to pose a threat to flow assurance. For information regarding the acceptance of reduced-bore ball valves in terms of pressure drop, process engineers are the most appropriate references. A second instance in which full-bore ball valves are required is when they are used before and after a pressure safety valve (PSV). A PSV is installed on pressure piping and equipment to release the overpressure fluid into flare lines. More information about PSVs is provided later in this chapter. In fact, PSVs are known as the ultimate safety solution in the event that other safety systems fail to work in order to release overpressure accumulation in pressurized equipment. The function of PSVs will be discussed in more detail later in this chapter. For most systems, more than one PSV is designed and selected; PSVs need to be maintained and calibrated, so if one is removed from the piping system, the other can be used as a backup. PSVs are located between two ball valves, one in the upstream direction (before) and one in the downstream direction (after). In operation, the ball valves installed before and after a PSV are always open. In spite of this, whenever a PSV is removed from a piping system for maintenance or modification, the ball valves before and after it must be closed. As a result, the backup PSV is used, and the ball valves located before and after the backup PSV should be opened. A full-bore ball valve should be installed before and after PSVs in order to facilitate the release of overpressure gas or liquid into the flare system (see Figure 3.4). A third reason the valve bore should be identical to the pipe bore is due to the operation of the piping injected gadget (PIG). There is a common practice of PIG running in pipelines, flowlines, and subsea manifold

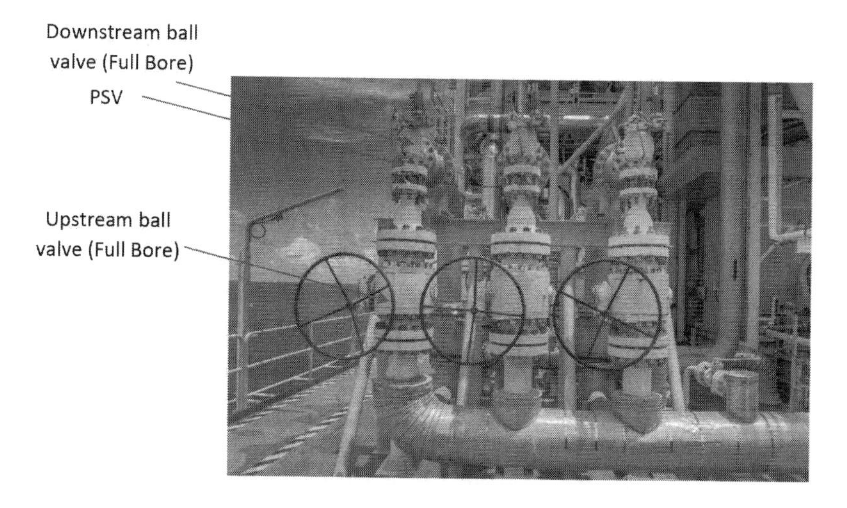

FIGURE 3.4 PSVs and full-bore ball valves on a compressor discharge line.

(Courtesy: Shutterstock)

headers. A PIG is a tool sent through a pipeline and moved forward by the pressure of the fluid inside. It is possible to send a PIG inside a pipeline for a variety of purposes, including cleaning and inspection. Pipeline valves must have the same internal diameter as the connected pipe in order to facilitate the passage of a PIG without obstruction. As a result, the ball valve is not considered to have a full or reduced bore but rather a ball valve with a special bore.

A number of factors should be considered when selecting the right ball valve for your application. Choosing between a top-entry and a side-entry ball valve is one of the most important decisions you will need to make. Each has its own set of unique features and benefits, so it is imperative that you be aware of their differences before making a decision. Taking a closer look at these two types of valves, let's compare them. Due to their ability to be installed from above, top-entry ball valves require a smaller installation area than side-entry ball valves. Due to this, they are ideal for use in tight spaces where there is limited access or access is difficult. A side-entry ball valve, on the other hand, must be installed from the side and typically requires more space than a top-entry valve. In general, side-entry design is more common. Figure 3.5 illustrates a side-entry ball valve located on an offshore platform. The valve can be assembled or disassembled from the side by unscrewing the side flange bolts.

Top-entry ball valves are typically equipped with a bonnet or cover that is bolted to the valve body. It is possible to gain access to the valve internals by unbolting the bonnet and body bolts from the top. As far as maintenance is concerned, it is possible to avoid removing top-entry valves from the line and provide online maintenance, whereas side-entry valves must be removed from the line in order to complete the work. Because of this, top-entry valves are often used in large pipelines or systems that need top access. Thus, the main advantage of

FIGURE 3.5 Side-entry ball valve in offshore.

(Courtesy: Shutterstock)

top-entry ball valves is that they can be serviced without having to remove them from the pipeline. This makes them easier and less time consuming to maintain, since workers or technicians only need to open up the valve body from above to access any internal components that may require replacement or repair. In order to perform any servicing or repairs on side-entry ball valves, they must be removed from the pipeline. Due to this, they are more labor intensive and require more downtime than top-entry valves. Top-entry ball valves are generally heavier than side-entry ball valves because of the heavy bonnet. Compared to side-entry valves, top-entry valves have a number of advantages. For example, they are more resistant to loads applied to the valve, such as those from connected piping or pipeline systems, and they are more flexible in the sense that they can be enlarged if higher strength is required or expected. It is common for top-entry ball valves to be welded to the connected piping, as opposed to side-entry ball valves that are typically attached via flanges at the ends. As a result of welding top-entry ball valves to the pipeline, flange connections are avoided and cost of flanges is reduced. The third option is a fully welded design, in which the body and bonnet are not bolted together. In contrast to side-entry or top-entry valves, which use gaskets to seal the joints between two bodies or bodies and bonnets, fully welded valves have a reduced possibility of leakage.

Another important feature of ball valves is the choice between floating and trunnion valves. Floating-ball valves have an unsupported ball, while trunnion-ball valves have a trunnion or support underneath the ball. Trunnion supports may take the form of plates or flanges. Generally, floating-ball valves are used for smaller and low pressure valves that require less force or torque to operate. The floating-ball valve is a type of valve that uses a floating ball to control the flow of liquid. Balls are connected to levers that can be used to open and close valves. When the valve is opened, the ball floats, allowing the liquid to flow through. Once the valve is closed, the ball sinks and prevents the liquid from flowing. There are many industrial and commercial applications for floating-ball valves. With this type of valve, the ball floats on top of the fluid, providing a tight seal while utilizing little force. Although floating-ball valves have many advantages, they also have some disadvantages. Floating-ball valves have the advantages of being able to handle a variety of fluid types, being easy to operate, being suitable for small spaces, and requiring little maintenance. While some fluids may cause the ball to stick, it may be difficult to repair, and it may not be suitable for applications involving high pressure and large size. In contrast, trunnion-mounted ball valves are designed to provide superior performance in critical applications. In order to meet the needs of different industries, trunnion-ball valves are available in different sizes and pressure ratings. Additionally, these valves are used in applications where high-quality valves are required that can withstand extreme temperatures and pressures. A trunnion-ball valve has a two-piece or three-piece body design that contains a ball with a hole in the middle. Balls are attached to trunnions, which allow the balls to rotate. As mentioned previously, the trunnion is attached to the body of the valve and is held in place by bearings. During the open position of the valve, the hole in the center of the ball aligns with the inlet and outlet ports of the valve. As a result, fluid can flow through the valve. When the valve is closed, the ball rotates so that the hole is no longer aligned with the ports, thereby blocking fluid flow. The following are some of the advantages of a trunnion-ball valve: trunnion-ball valves are very sturdy and can withstand high pressures. They are easy to operate and can be opened and closed in a short period of time. A trunnion-ball valve provides a good seal and can be used with a variety of liquids and gases. Trunnion-ball valves can be expensive, which is one of the disadvantages associated with this type of ball valve. If they break, they can be difficult to repair. In some applications, trunnion-ball valves can be difficult to install due to their large size and bulk applications.

Some ball valves have special designs. The first type of ball valve with a special design is a V-notch ball valve. The V-notch ball valve is designed with a V-shaped ball, as illustrated in Figure 3.6, and is designed to control flow effectively with a low risk of cavitation, noise, corrosion, and vibration, the most common problems associated with standard ball valves. V-notch design allows precise flow, modulating and controlling flow with 30°, 60°, or 90° V-notches. V-notch ball valves are designed to control straight-through flow in a wide range of fluids with high capacity. Their use is particularly beneficial in the processing of pulp and paper stocks as well as in the processing of liquids containing suspended solids. This type of ball valve has a small operating torque and can be used in a wide range of applications. A C-type ball valve is another type of ball valve with a special design that can be used to control flow. The ball in this valve is C shaped (see Figure 3.7), as implied by its name. In an eccentric hemisphere valve, also known as a C-ball valve, the hemisphere rotates to open or close the valve.

Three-way or three-way ball valves are special valves with multi-ports, that is, three or four ports, that are typically used for diverting or mixing the flow. Three-way ball valves may be selected and operated in different sectors of the oil

FIGURE 3.6 V-notch ball valve.

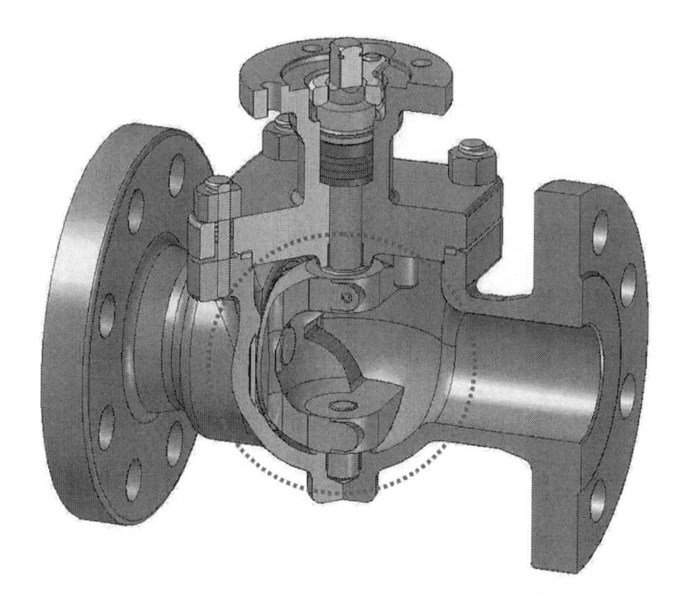

FIGURE 3.7 C-type ball valve.

FIGURE 3.8 Three-way ball valves.

and gas industry. Depending on the application, this type of valve may be designated with an L-port or a T-port (see Figure 3.8). The number of ports could be more than three, as in the case of a four-way ball valve, or even higher. Ports can be manufactured by the valve supplier upon request by the purchaser. For flow switching, four-way valves and other multi-bore valves are ideal.

3.2.2 Gate Valves

3.2.2.1 Through Conduit Gate Valves

Through conduit gate valves are used to turn on and off a fluid or to start and stop it. As with ball valves, TCG valves are not recommended for fluid control or intermediate positions, which may result in severe erosion and wear inside the valve. In sandy or particle-containing services, TCG valves are a better choice than ball valves. In contrast to gate valves, ball valves have a number of advantages. The ball valve is easier and faster to open and close than the gate valve, requiring only a 90° rotation of the ball. It is possible for gate valves to have a greater height than ball valves, especially in large sizes of 12" and above. When there is a limitation on vertical space, this additional height could pose a problem. Furthermore, it may be difficult to gain access to the top of a large gate valve for operation, so a platform may be required to provide the access needed. Another disadvantage of a gate valve over a ball valve is the friction between the stem and its sealing, which is higher in a gate valve than a ball valve due to the linear movement of the stem. In a gate valve, a higher friction between the stem and stem sealing results in the stem seals wearing out and eventually causing damage to the stem. As with ball valves, TCG valves may be full bore or reduced bore. Since these valves are used in dirty environments, full-bore TCG valves are the preferred choice. It is easier for particles and dirt to pass through a valve with a full bore. The stem and closure members (disks) of a gate valve are linear in motion, unlike ball valves, which rotate 90°. As with standard ball valves, all gate valves have two seats. As a standard design, gate valve seats are metallic. There is a bonnet or cover on top of the body of the TCG valve, which signifies that it is a top-entry valve. In most cases, gate valves consist of a body and a bonnet that are bolted together. It is possible to access the internals of the valves from the top for maintenance or other purposes by unscrewing the bolts that secure the body and bonnet together. Gate valves have a top-entry design that provides access to the valve internals when the bonnet is removed. Top-entry design has some advantages, such as online maintenance, meaning that there is no need to dismantle the valve from the piping for maintenance. TCG valves must be equipped with a secondary stem seal, also known as a backseat bushing. The backseat is a portion of the valve's inner bonnet where the stem is located and prevents fluid from flowing to the stem sealing area. To provide fluid isolation between the stem sealing areas, a conical area sits on the backseat area. Due to their high bodies and upward stem and gate movement, gate valves have a greater height than ball valves. It is possible to consider the height of a gate valve a disadvantage. As a general rule, TCG valves (both slab and expanding) are heavier and bulkier than ball valves, but this is not always the case. There is no standard weight for valves; their weight is determined by the manufacturer of the valve. Ball valves typically operate at a faster rate than TCG valves. The wear on the stem is greater for a gate valve than a ball valve, since the stem moves up and down and is in contact with all of the sealing and packing on the stem. The stem of a ball valve rotates only 90°, so there is much less contact area between the stem and the seal. Unlike ball valves, it is not proposed to use metallic stem sealing for gate valves, as it can wear the stem significantly. There are two types of TCG valves: slab gate valves and expanding gate valves.

3.2.2.1.1 Slab Gate Valves

In process services such as hydrocarbon (oil and gas), slab gate valves are used to stop and start the flow of fluids. The opening and closing of the valve are carried out through a linear movement of the stem that transfers to the gate (disk), also called the closure member. Gates are equipped with fully circular port openings that are capable of aligning with the flow in open positions or blocking the flow when closed. Between two seats, the gate moves or slides. In a standard design with two floating seats, the downstream seat and the disk are in contact with metal to metal (MtM). It is the seats that ensure that the body and gate are in continuous contact with one another. Through fluid pressure, the disk is pushed to the downstream seat. From a sealing point of view, this type of slab gate design is known as a pressure energized–upstream downstream sealing only. The sealing between the downstream seat and disk is a function of both spring force and fluid pressure. When the internal fluid service is at low pressure, it is important that the valve be capable of providing acceptable sealing. Another important consideration is that both seats of a slab gate valve should be floating type, which means that they are energized by springs. In contrast, a design with a fixed seat is not recommended because the downstream seat cannot be sealed tightly. Generally, slab gate valves are bidirectional, meaning that they can be rotated 180° and attached to piping in either direction. There is no doubt that the gate is the most expensive component of a slab gate valve. It should be polished perfectly and coated with a thin layer of tungsten carbide in order to prevent wear due to erosive fluid and galling due to friction. In slab gate valves, the stem can be driven by an actuator or manually by a handwheel. Stem sealings are used to provide tightness between the stem and the body and bonnet. Figure 3.9 shows a slab gate valve.

3.2.2.1.2 Expanding Gate Valves

Through conduit expanding gate valves, also known as expanding gate valves, are designed, selected, and operated to isolate the piping system or components when they are closed. Similarly to slab gate valves, expanding gate valves should not be used for flow control (throttling) since their internal components would be subject to vibration, excessive wear, and damage. The main components of an expanding gate valve are the body, the bonnet, the seats, and the disk. In Figure 3.10, the main components of an expanding gate valve are illustrated. All components of the expanding gate valve are similar to those of the slab gate valve, with the exception of the disk and sealing mechanism. As opposed to slab gate valve seat sealing, which is achieved with floating seats and fluid pressure that pushes the disk downstream, an expanding gate valve is a torque-sealed valve, which means the stem force is transmitted to half disks, which are pushed strongly toward the seats by the stem force. When a double expandable gate valve is closed or open,

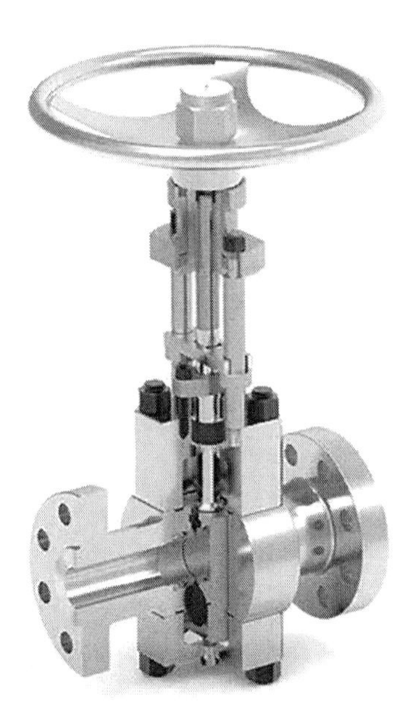

FIGURE 3.9 A slab gate valve.

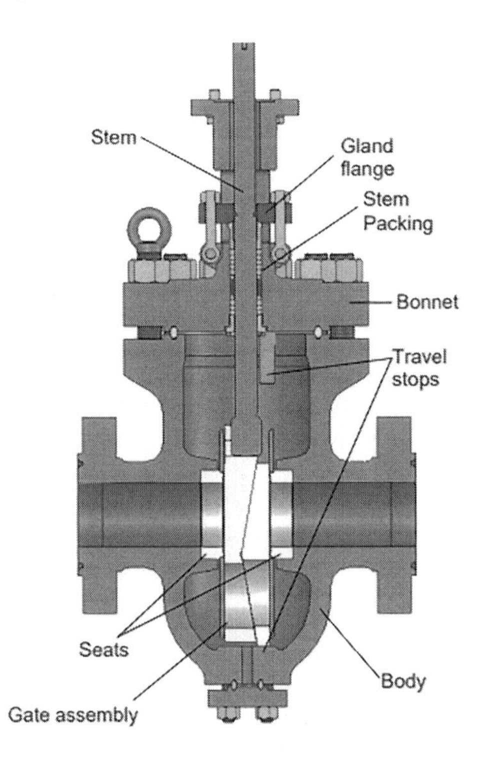

FIGURE 3.10 Expanding gate valve.

two half disks are pushed and expanded strongly against both seats. It is important to note that strong sealing between the disk halves and seats will only be achieved in the closed position of a single expanding gate valve.

There is a significant difference between the mechanism of an expanding gate valve and that of a slab gate valve in terms of the closure member or disk. In a double expanding gate valve, the gate is composed of two parts: one female and one male, with a sealing or lever between them (see Figure 3.11). A half disk slides against another half disk. Using a T-shaped slot connection, the stem is connected to the top part of the female section. A T-shaped connection between the stem and disk provides an anti-blowout feature for the stem, as well as the ability to move the stem sideways with the disk. By applying stem force to the contact surfaces between the male and female halves, the disks expand laterally. The entire disk assembly is placed between two seats with or without a guided ring or seat skirt. The female section drags the male disk, and the entire disk assembly moves upwards and downwards between the upper and lower travel limits. At both ends of the disk assembly, travel stops restrict movement, and the stem force is transferred to the female part of the disk, resulting in a wedging effect. The wedging effect can be defined as the process of applying force to the thick end of a wedge on top, which then transfers the force to the objects located along both sides of the slope (see Figure 3.12). Accordingly, the mechanical force of the stem is transferred to the female disk, which causes the disk to expand toward both seats.

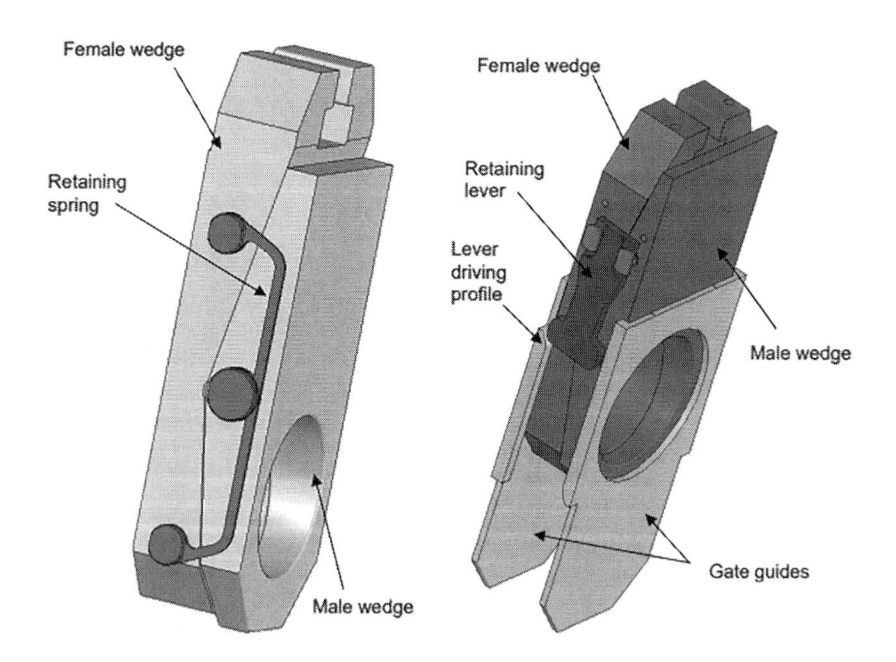

FIGURE 3.11 Male and female disk segments.

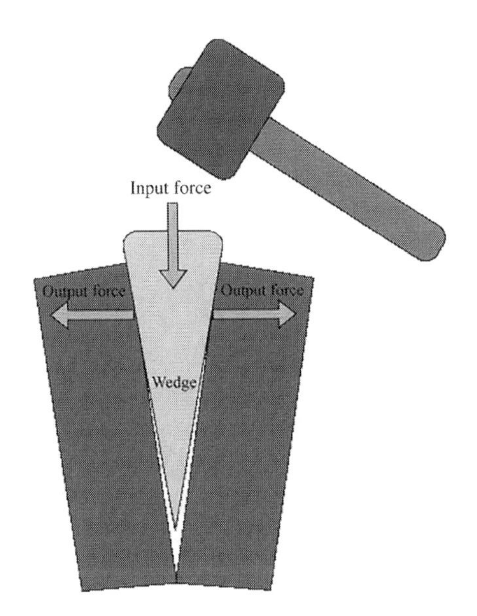

FIGURE 3.12 Wedging effect.

3.2.2.2 Wedge Gate Valves

As with other types of gate valves (slab, expanding), wedge gate valves are used for on/off functions and for flow isolation. A wedge-shaped gate valve has a gate in the shape of a wedge. It is not recommended to use a wedge gate valve for throttling or fluid control, since it can cause excessive wear and erosion to the internals of the valve, such as the disk, seats, and lower part of the stem. It is also important to note that keeping a wedge gate valve in a semi-open position causes the disk to chatter. It is important to note that wedge gate valves, like expanding gate valves, are torque-seated valves, which means that the seal between the valve disk and seats is achieved by stem force rather than fluid pressure. Wedge gate valves share the disadvantages mentioned earlier in this chapter for gate valves in comparison to ball valves, such as their height, slower speed of operation, and greater stem sealing wear. As opposed to TCG valves, wedge gate valves have other disadvantages; they are not suitable for particle-containing services due to the possibility of sand and particles accumulating at the bottom of the wedge and damaging it during closure or preventing it from closing completely. In Figure 3.13, a solid wedge gate valve is shown in its closed position on the right, and the wedge of the valve is shown on the left. As the handwheel is rotated counter-clockwise, the stem and wedge are lifted, allowing the valve to open and allow fluid to pass through.

Figure 3.14 illustrates a wedge gate valve and its component parts. The body and bonnet are connected by bolts. There is a type of seal called a gasket that is placed between the body and bonnet in order to prevent leaks from occurring. The packing is the stem sealing, which is usually made of graphite. The gland is

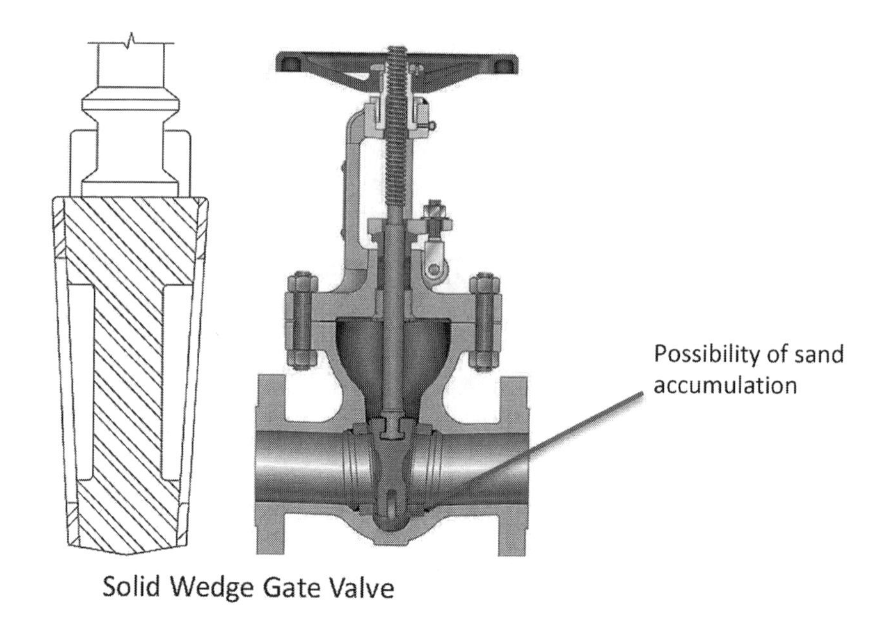

FIGURE 3.13 Wedge gate valve.

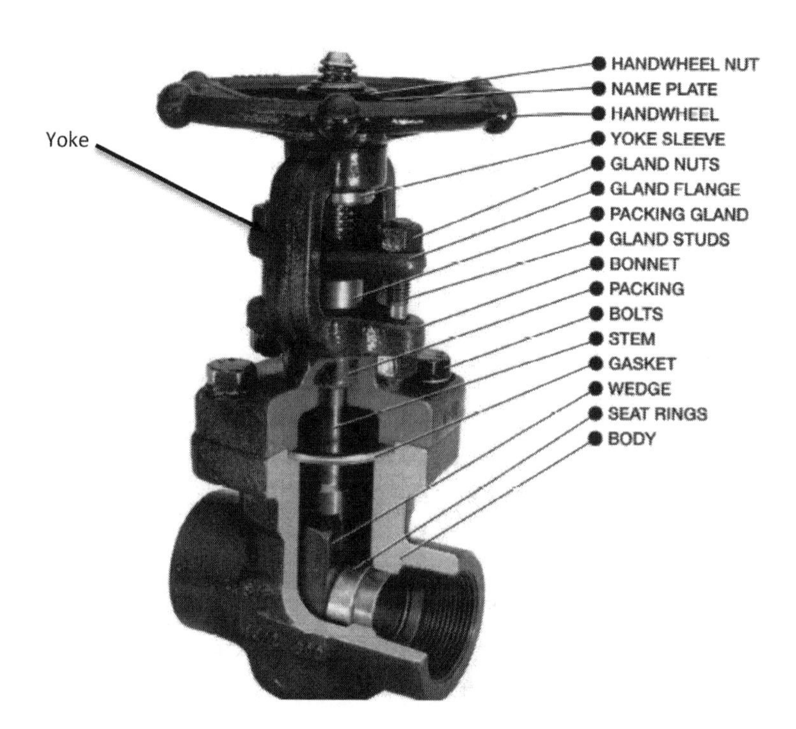

FIGURE 3.14 Wedge gate valve with parts list.

located at the top of the packing and below the gland flange. Bolts and nuts are used to secure the gland flange, which provides axial force to the gland and packing. As a result of the axial force, the packing expands radially and provides an effective seal around the stem. It is very important to seal the stem since any leakage from the stem directly enters the environment. To connect the handwheel or actuator to the valve bonnet, the yoke is integrated with the valve bonnet. A yoke sleeve, also known as a yoke nut, stem nut, or yoke bushing, is located around the stem and transfers the load from the handwheel to the stem.

During the opening and closing of a wedge gate valve, the disk or gate moves perpendicularly to the flow direction. In addition, the wedge gate valve can only be fully opened and fully closed and cannot be adjusted or throttled in any way. There are two sealing surfaces on the wedge gate valve, which are usually arranged in the form of a wedge. In general, the wedge angle varies depending on the valve parameters and is usually 5°. It is possible to integrate the gate of the wedge gate valve, which is known as a solid gate. As an alternative, it can be made into a gate that can produce trace deformation to improve its manufacturing capabilities and compensate for any deviation in the sealing surface angle during processing, which is called a flexible gate. There is a linear movement between the gate and stem of the wedge gate valve. The majority of wedge gate valves are forced seal valves. It is necessary to apply an external force to the gate when the valve is closed in order to ensure that the sealing surface of the valve is sealed.

3.2.3 Plug Valves

The plug valve (see Figure 3.15) is another type of valve that may be used for the isolation of fluids or the control of fluid flow. Plug valves can also be used to throttle fluids in a limited manner. Plug valves are recommended for dirty or particle-containing services. Similar to ball valves, plug valves move between open and closed positions by rotating the steam and plug (closure member) 90°. The plug consists of a cylindrical or conical tapered shape with a hole inside. In the case of a plug with a hole aligned with the flow (see Figure 3.15), the valve is open and the flow passes through it. After rotating the plug 90°, the hole inside the plug is perpendicular to the direction of the flow, which causes the solid part of the plug to be placed in front of the flow, which closes the valve. It is common for plug valves to make metal-to-metal contact between the plug and the body, but lubricant can sometimes be injected onto the plug to improve sealing capability and to reduce friction between metal and metal. Plug valves that are lubricated are referred to as lubricated plug valves. The plug valve, as well as the TCG valve, is an excellent choice for fluids containing particles. Plug valves have the advantage of being more compact than TCG valves, generally speaking. It should be noted, however, that plug valves require lubricants for sealing, which is not desirable from an operational standpoint. It is recommended that the lubricator be injected into the plug valve at a specific interval. There are non-lubricant plug valve designs available. For example, Teflon-sleeved plug valves have a Teflon sleeve surrounding the plug and fitting between the plug and body. It should be

FIGURE 3.15 A plug valve.

noted, however, that Teflon is a soft (non-metallic) material and is not suitable for applications that contain particles. Thus, Teflon- or PTFE-sleeve plug valves are not suitable for dirty services due to the possibility of damage to the PTFE sleeve.

Additionally, a plug valve may have more than two ports. The flow from one port of a three-way plug valve can be directed to either the second or third port. In addition to switching flow between ports 1 and 2, 2 and 3, or 1 and 3, a three-way plug valve may also connect all three ports together. The flow-directing possibilities of multiport plug valves are similar to those of multiport ball valves. There is also the option of having one port on one side and two ports on the other side of the plug valve.

The advantages of plug valves include the following: It has a simple design and fewer parts; it can be quickly opened or closed; and it offers minimal flow resistance. By using multi-port designs, the number of valves required can be reduced and the flow direction can be changed. In addition, it provides a leak-tight service. In addition to being easy to clean, their bodies can be removed from the piping system without causing damage. In comparison with ball valves, plug valves have some disadvantages, such as higher cost and greater pressure drop.

3.2.4 Butterfly Valves

A butterfly valve is another type of valve that is used for the isolation of flow or for on/off operation (see Figure 3.16). It is also important to note that butterfly valves can be used for controlling or regulating flow. Butterfly valves are quarter-turn valves that can be moved between an open and closed position by rotating the stem and closure member (disk) 90°. Butterfly valves are easy to operate due to their quarter-turn design. A butterfly valve requires less torque or force to operate than a ball valve, since the disk of a butterfly valve is relatively light

FIGURE 3.16 Flanged connection butterfly valve.

(Courtesy: Shutterstock)

in comparison to the ball of a ball valve. Butterfly valves have the advantage of being lighter, cheaper, and smaller than ball, gate, and plug valves. Additionally, butterfly valves can also be designed and manufactured in a wafer type design (flangeless body), which significantly reduces the valve's weight. Wafer designs are flangeless designs with facings that permit installation between ASME and manufacturer standard flanges (MSS SP). It is noteworthy that the weight of valves, including ball and butterfly valves, varies from one valve supplier to another. The face-to-face values of ball and butterfly valves are usually determined by international standards such as ASME and API. As a result of the aforementioned advantages, wafer-type butterfly valves are widely used for utility services (non-aggressive fluids) such as water, seawater, and oxygen in the offshore industry instead of ball valves.

Compared to other types of valves, butterfly valves have a lower maintenance cost due to their simple design and fewer moving parts. Due to the reduced bore of butterfly valves, they create a greater pressure drop in the piping system than ball and gate valves. The use of butterfly valves offshore should be limited to utilities such as air and water in low pressure classes, as butterfly valves are not as robust as ball and TCG valves. ASME B16.34, the standard for valves, defines low pressure classes as CL150 and CL300, which correspond to 20 and 50 bar, respectively. Butterflies have a simple working principle; the handwheel, lever, or actuator transfers the load to the valve stem, which rotates. Consequently, the rotation of the stem is transferred to the disk inside the valve. Both the disk and stem rotate together when the disk is in an open position and the stem is in a closed position.

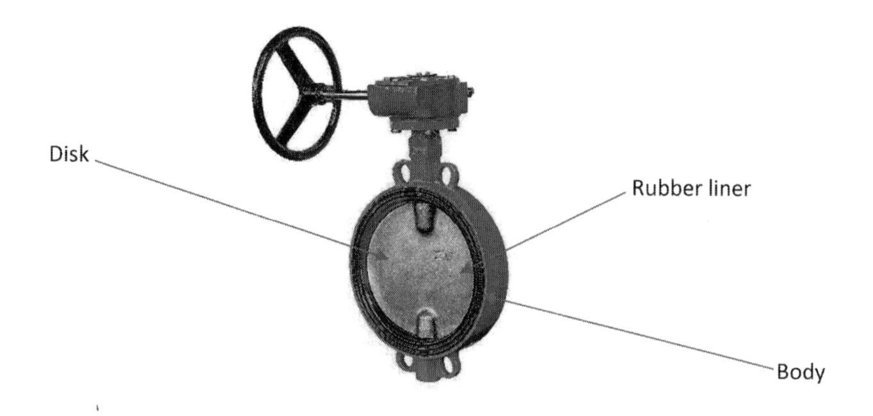

FIGURE 3.17 Rubber-lined (concentric) butterfly valve.

(Courtesy: Shutterstock)

As noted above, butterfly valves may be used to control or regulate flow, in which case the disk is positioned in an intermediate position between open and closed.

The design of butterfly valves can be divided into two main categories: concentric and eccentric. There can be a double or triple offset on eccentric butterfly valves. Typically, concentrator butterfly valves are rubber lined, which means that the valve body is fully covered by a layer of rubber, as illustrated in Figure 3.17; the rubber liner serves to seal the disk against the body. Butterfly valves with a concentric design do not have any offset. When it comes to butterfly valves, what does the term "offset" mean? Eccentric butterfly valves can be offset in three different ways. The first offset refers to the position of the disk in relation to the body. In concentric butterfly valves, the disk is located in the center of the body, whereas in eccentric butterfly valves, the disk is located off to one side of the body. In addition, there is an offset related to the position of the stem in relation to the disk. A concentric butterfly valve has a stem that passes through the middle of the disk exactly. The stem of a double or triple offset butterfly valve, however, does not pass through the middle of the disk. The third and last offset relates to the offset between the seat and the disk during sealing. The first and second offsets are present in a double offset butterfly valve, whereas the third offset is present in a triple offset butterfly valve. In general, double and triple offset butterfly valves are more robust than lined or concentric butterfly valves. The primary weakness of concentric butterfly valves is their liner, which may be damaged by particles, frequent cycling (opening and closing), or operation at high differential pressures. All three butterfly valve offsets are illustrated in Figure 3.18.

3.2.5 Modular or Combination Valves

It is possible to integrate two ball valves in a single body with a needle valve separating them. These valves are referred to as modular valves, as illustrated

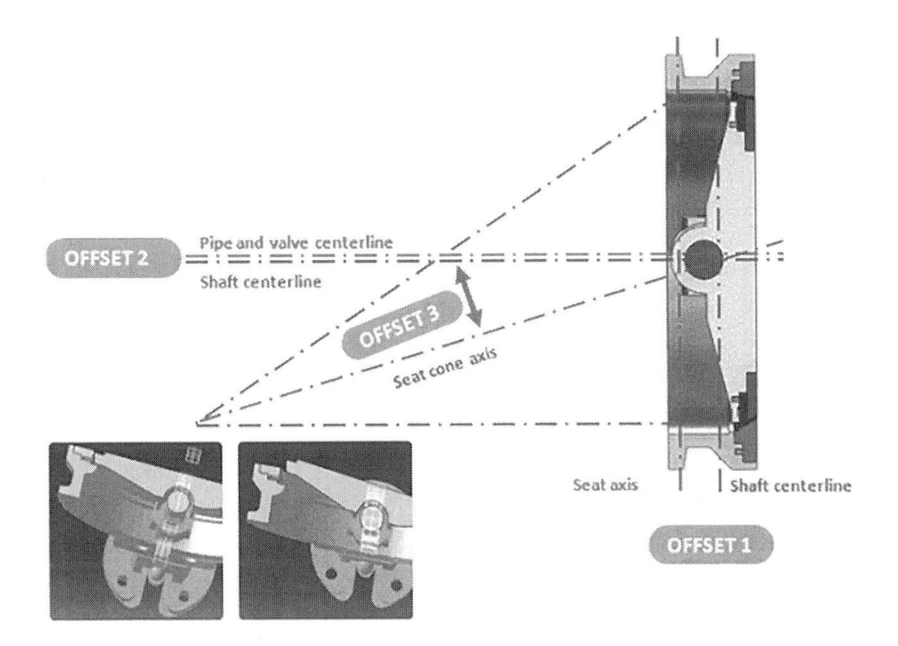

FIGURE 3.18 Butterfly valve offsets.

FIGURE 3.19 Modular valve including two ball valves and a needle between them.

in Figure 3.19. This type of valve is sometimes called a double block and bleed (DBB) valve, as it provides double isolation through the use of two balls and a needle valve between the two balls that acts as a bleeder. It is possible to use this type of valve on chemical injection lines to provide a double barrier between the main process line and the chemical injection line. Two blockages are provided by two ball valves, which provide increased safety and reliability. If one valve fails to provide isolation, the other may provide it as a backup. The integration of all three valves into a single body has several advantages, including reduced cost, weight, and space. It is the primary purpose of modular valves to provide double isolation by closing two ball valves at the same time. To improve safety and reliability, double isolation may be necessary in high pressure class piping and/or aggressive fluids. The use of two blocks for isolation rather than one increases the cost, weight, and space requirements of the valve as well as the complexity of its design. The needle valve, located between the two ball valves, acts as a bleeder, releasing the fluid trapped between them. Modular valves are very common in small sizes, that is, less than 2". The important thing to remember is that a modular valve containing two valves cannot be used for fluid control.

3.3 FLUID CONTROL VALVES

Some cases require that the amount of flow passing through the piping be controlled. Flow control can be used to adjust key process variables, such as pressure, temperature, and fluid level. Valves for fluid flow regulation are designed to regulate the flow by changing the fluid passage and maintaining essential process variables as close as possible to the desired set point. Fluid control can be achieved using butterfly valves, as discussed previously. Further, globe, axial control, needle, and choke valves are suitable for fluid regulation.

3.3.1 Globe Valves

Globe valves have been widely used in different sectors of the oil and gas industry. Globe valves are covered mainly by American Petroleum Standard (API 602), titled "Gate, Globe, and Check Valves for Sizes DN 100 (4" Size) for the Petroleum and Natural Gas Industries." Other standards include BS 1873, "Specification for Steel Globe and Globe Stop and Check Valve (Flanged and Butt-Welding Ends) for Petroleum, Petrochemical and Allied Industries," as well as ISO 15761 "Steel Gate, Globe and Check Valves for Sizes DN 100 and Smaller, for the Petroleum and Natural Gas Industries." A globe valve, different from ball valve, is a type of valve used for regulating flow in a pipeline, consisting of a movable plug or disc element and a stationary ring seat in a generally spherical body. A globe valve is a linear motion valve and is primarily designed to stop, start, and regulate flow. The disk of a globe valve can be totally removed from the flow path, or it can completely close the flow path. The American Petroleum Institute Recommended Practice (RP) 615 valve selection guide specifies that globe valves may be used to block fluids, but based on industrial experience, the author recommends avoiding the use of globe valves to stop or start fluids. Globe valves can, however, be used in fluid control applications when 100% flow passage is required during complete opening or 0% flow passage during complete closing. One of the most common types of globe valves is a T-pattern or standard design, while the other is a Y-pattern design. In Figure 3.20, a globe valve with a parts list is shown. In Figure 3.21, a globe valve with an actuator for automatic operation is referred to as a control valve.

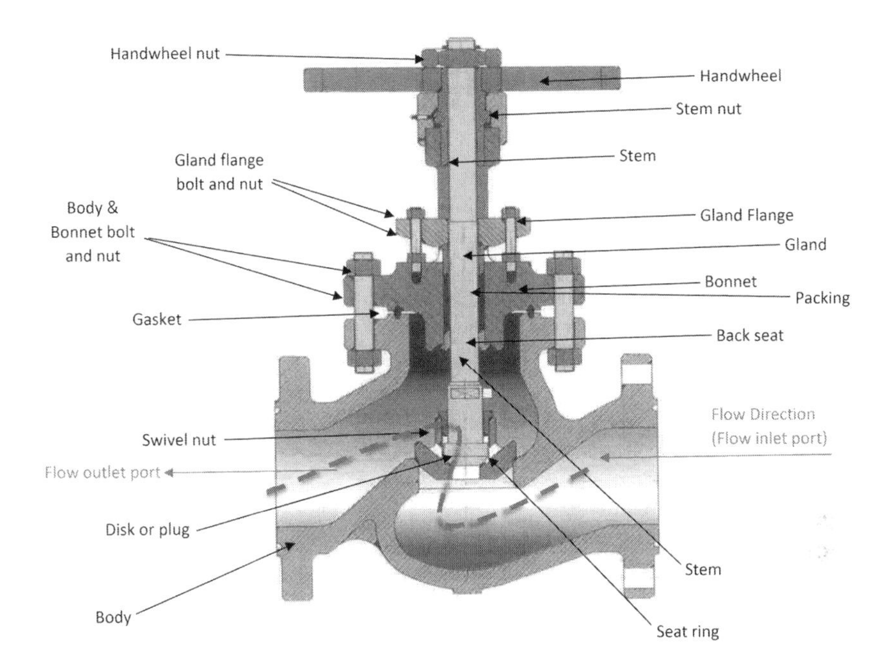

FIGURE 3.20 Globe valve with parts list.

FIGURE 3.21 Control valve including a globe valve with diaphragm actuator.

(Courtesy: Shutterstock)

As illustrated in Figure 3.20, the globe valve is composed of a body that is connected to the bonnet or cover by bolts and nuts, and a gasket is placed between the body and bonnet for sealing purposes. In the closed position, the disk or plug rests on the seat. Connecting the stem to the plug or disk requires the use of a disk swivel nut or disk nut. If a disk nut is used on a rotating stem globe valve, the stem's rotation is not transferred to the disk. Using a disk nut ensures that the stem and disk are not firmly connected, ensuring that the disk is always positioned correctly on the seats without any rotation of the plug. Prior to packing, the back seat of the valve stem is considered the primary sealing area. In the figure, the valve is shown closed. The operator opens the valve by rotating the handwheel counter-clockwise and transmitting the force through the stem nut to the stem. The stem and disk connected to the valve move upward, and the valve opens. In this way, the fluid passes through the valve, following the route indicated by the dotted lines within the valve. A fluid enters the valve from the right side, and it leaves the valve after two 90° rotations near the seat and plug. As shown in the figure, a globe valve does not have a bidirectional flow direction, because the flow direction is coming from under the disk, seat, and stem area. There is a significant pressure drop inside the valve as a result of the two 90° rotations of the fluid.

As a result of the significant pressure drops, the bubbles in the liquid fluid service are vaporized, which is known as flashing. Following passage through the pressure drop area close to the disk, the bubbles will regain their pressure and burst. Cavitation, or the bursting of bubbles, is one of the most common operational problems associated with T-pattern or standard globe valves. Normally, the body, seat, plug, and stem of standard globe valves are severely damaged by cavitation, which can be considered a type of erosion/corrosion. Noise and vibration are also negative outcomes of cavitation. Cavitation can cause irregular pitting and erosion in the trim (seat and plug) and body of globe valves, as well as downstream piping. Cavitation damage appears in the form of small pits, which are very similar to corrosion damage in the plugs of globe valves. As a result of cavitation, corrosion is intensified. This could be called "cavitation corrosion." Cavitation mitigation techniques include increasing the diameter of the stem, strengthening the connection between the stem and plug, and applying hard-faced materials such as Stellite 6 or 21 to the seat and plug. In valves, Stellite is a type of cobalt and chromium alloy used for wear resistance in parts such as the seat and plug. Keeping the valve less than 20% open is not recommended from an operational perspective, as this can cause severe wear and erosion, which can intensify the cavitation effect. Another solution is to replace the valve with either a Y-type globe or an axial control valve. A Y-type or Y-pattern globe valve, as illustrated in Figures 3.22 and 3.23, reduces the pressure drop inside the valve by reducing the rotation of flow inside the valve to a maximum of two times and changing the flow direction at a maximum of 45° rather than 90° in each rotation. A reduction in pressure drop results in a reduction in cavitation.

The American Petroleum Institute released the first edition of its new globe valve design standard (API 623) in 2013 in order to prevent and control problems associated with globe valve operation such as cavitation, vibration, and leakage. According to API standard 623, "Steel Globe Valves—Flanged and Buttweld

FIGURE 3.22 Y-type globe valve from outside.

(Courtesy: Shutterstock)

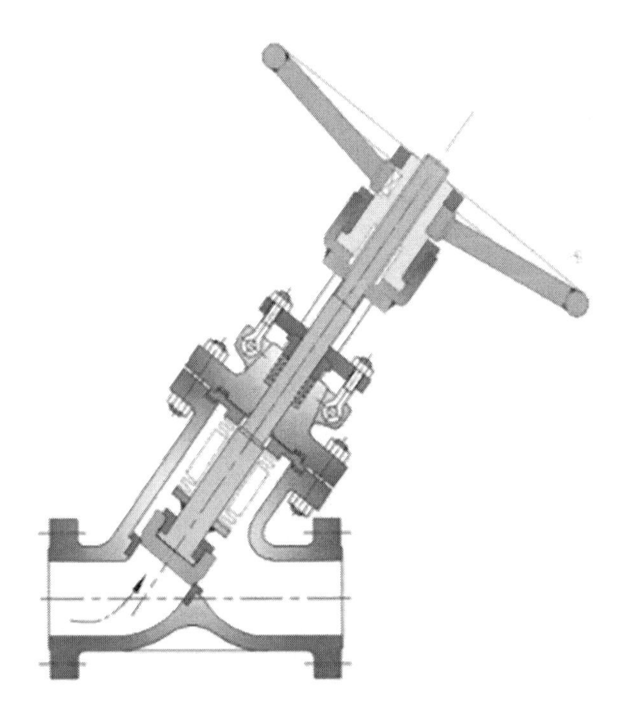

FIGURE 3.23 Y-type globe valve internals.

Ends, Bolted Bonnets" are covered by the following features: bolted bonnet, external screw and yoke, rotating rising and non-rising stems, rising and non-rising handwheels, Y-pattern globe valves from 2" to 24" and pressure classes of 150 (pressure nominal [PN] = 20 bar), 300 (PN = 50 bar), 600 (PN = 100 bar), 900

(PN = 150 bar), 1,500 (PN = 250 bar), and 2,500 (PN = 420 bar). Especially for high pressure classes, API 623 specifies hard facing on the seat, plug, and guided disk. The stem diameter in API 623 follows the principles of the API 600 "Cast Steel Gate Valves Standard" with different values. The stem diameter values in API 623 are larger than other globe valve standards like BS 1873 to avoid stem and plug separation (breaking).

3.3.2 Axial Valves

In comparison with globe valves, axial control valves are known as the best valves for flow control applications. An axial control valve is characterized by a smooth, streamlined flow path inside the valve, which provides turbulence-free flow without operational problems such as vibrations and noise. This valve has a low pressure drop, which prevents cavitation and flow loss inside it. Additionally, this valve is compact and lightweight, which is advantageous. Axial control valves have a light disk and a short distance between the disk and the seat, making the valve suitable for fast opening and closing. Furthermore, the disk does not slam against the seat, so this type of valve is non-slamming. The axial valve in Figure 3.24 is in the open position. In order to close the valve, the stem should be moved downward in order to push the disk forward. Once the disk reaches the seat, the valve will be closed, and the fluid will cease to flow. In order to control the flow of fluid through the valve, the stem movement can be adjusted to keep the disk in an intermediate position. Additionally, a cage is installed between the disk and the seat of the valve that affects its flow characteristics.

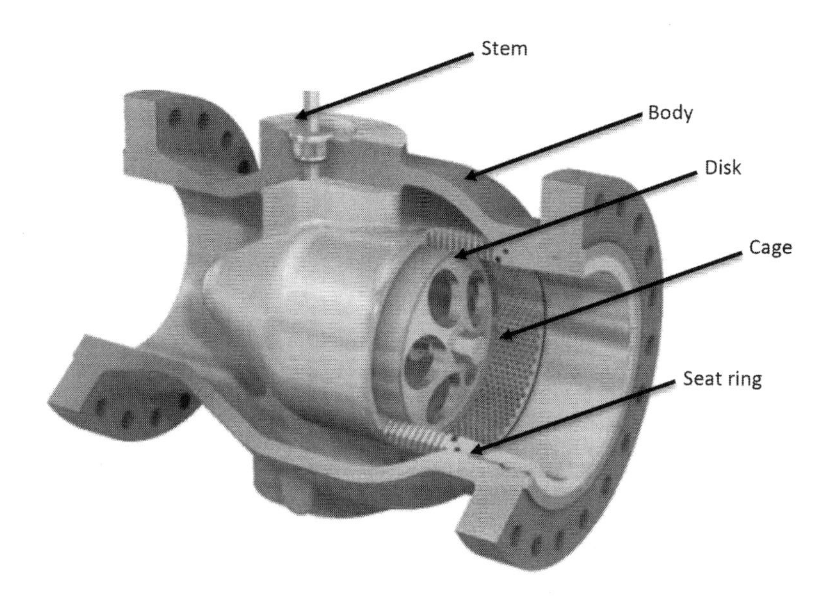

FIGURE 3.24 An axial valve.

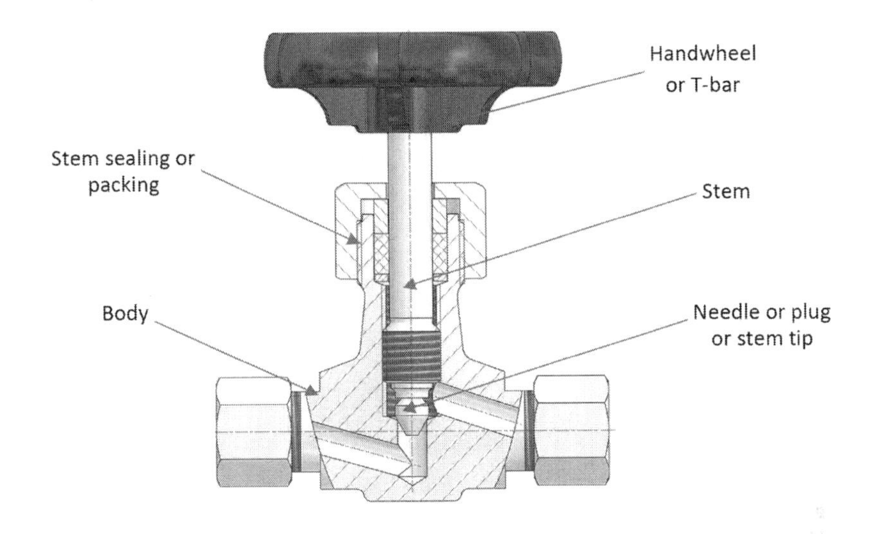

FIGURE 3.25 A needle valve with integrated stem and plug.

3.3.3 Needle Valves

A needle valve has a small port and a needle-shaped plunger that is threaded. Although it is generally only capable of relatively low flow rates, it allows precise control of the flow. A needle valve uses a tapered needle or plug to gradually open a space to allow fine control of flow. Through the use of a stem, it is possible to control and regulate the flow of fluid. A needle valve consists of a relatively small orifice with a long, tapered seat and a needle-shaped plunger or plug that fits perfectly in the seat. Typically, the plug (needle) and stem of the valve are integrated, as illustrated in Figure 3.25.

In Figure 3.26, a needle valve is illustrated that is used to isolate a pressure gauge from high-pressure fluid service when the gauge is removed for maintenance or calibration. It is not recommended to select a needle valve for isolation, especially for high pressure class services. Whenever high pressure fluid services are to be isolated, modular valves are the best and safest choice of valve.

3.3.4 Choke Valves

Choke valves are another type of control valve installed on production wellheads to control the flow being produced. Also, choke valves are capable of stopping production if something goes wrong downstream. After the choke valve, downstream refers to the piping and separator. A choke valve is shown in Figure 3.27 (left) on the right side of some Christmas tree valves. In the oil and gas industry, a Christmas tree is an assembly of valves, spools, and fittings used to control the flow of oil from a well. In the right hand corner of the picture, you can see a choke valve mounted on a Christmas tree. In Figure 3.27, the fluid enters the choke

FIGURE 3.26 A needle valve for pressure gauge isolation in high pressure fluid application (not recommended).

(Courtesy: Shutterstock)

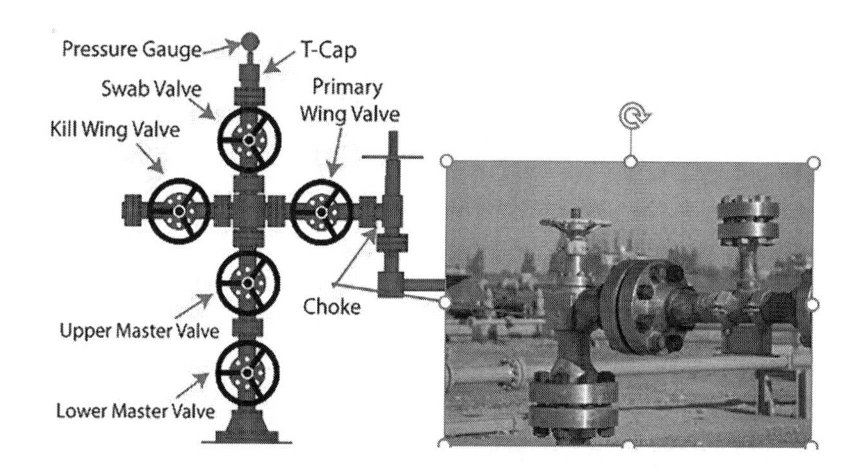

FIGURE 3.27 Choke valve on a wellhead.

(Courtesy: Shutterstock)

valve from the bottom part, inside the vertical line; after rotating 90° inside the valve, the fluid exits the valve along the horizontal line on the right-hand side. Due to the 90° rotation of the fluid inside choke valves and the resulting pressure drop, choke valves are at high risk of operational problems such as erosion and cavitation. Figure 3.27 shows a choke valve with a stem and handwheel on the right side. In order to adjust the flow passage, the handwheel can be used to move the stem and connected plug upward or downward. A counterclockwise rotation of the handwheel causes the stem and connected plug to move upward, which increases the opening and allows more flow to pass through the valve. Rotating the handwheel clockwise closes the valve in the opposite direction. An actuator can operate a choke valve automatically; in that case, there is no handwheel. To conclude, choke valves are flow control devices used primarily in the oil and gas industry during the production of crude oil from an oil well. A valve such as this is typically used to control the flow rate of drilling fluid, such as crude oil, in a high-pressure environment. Choke valves also serve the purpose of reducing downstream pressure in flow pipes by creating pressure drops across the valve.

3.4 NON-RETURN VALVES

A non-return valve, also known as a check valve, is used to prevent the backflow or reverse flow of liquids or gases. In order to keep the valve open and allow fluid to pass, check valves are marked with the direction of flow on the body of the valve. Check valves do not have any operating mechanism; they open as a result of fluid pressure. In the event that the fluid stops and/or returns, the valve closes and prevents the flow from moving in the opposite direction. A check valve is an automatic valve that opens when forward flow is present and closes when reverse flow is present. Piston check valves, swing check valves, dual-plate check valves, and axial check valves are among the different types of check valves. The following is an explanation of the different types of check valves used in the oil and gas industry.

3.4.1 SWING CHECK VALVES

In swing check valves, a disk swings around a hinge. There is a pin or shaft that passes inside the upper part of the hinge that connects the hinge to the body of the valve. On the left side of Figures 3.28 and 3.29, a swing check valve is shown in its closed position. When the valve closes and fluid returns to the upstream side (before) the disk, the closed disk prevents the fluid from flowing back. The disk is completely seated on the seat when it is closed. On the right side of the figure, the

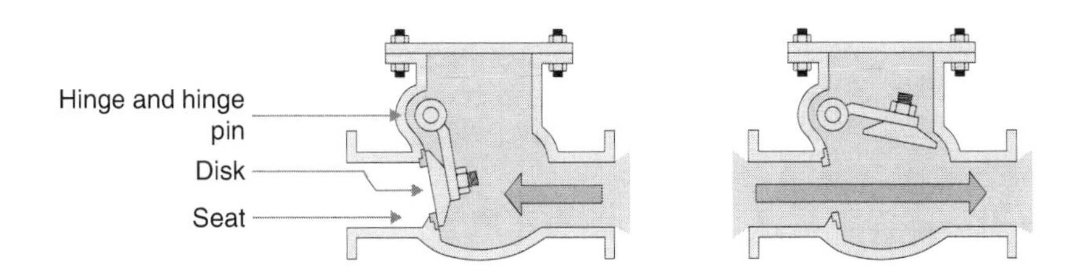

FIGURE 3.28 Swing check valve in open position (right side) and closed position (left side).

(Courtesy: Shutterstock)

FIGURE 3.29 Swing check valve in closed position.

(Courtesy: Shutterstock)

fluid is moving from upstream of the valve, which allows the fluid to flow through the valve. Once the flow of fluid has stopped, gravity and the weight of the disk cause it to return to its initial position on the seat, closing the valve. Thus, gravity and reverse flow both contribute to keeping the disk tight to the seat. It is important to understand that both of these parameters could result in the disk slamming against the seat during shutdown. The slamming of swing check valves is one of the most common operational problems. In addition to damaging the disk and wearing out the hinge pin, slamming can have a variety of negative effects. As a result of slamming, pressure waves can build in the pipes and cause water hammering. A high level of noise and acoustic fatigue may occur as a result of water hammering, which can damage pipes, valves, and instrumentation. To prevent slamming and other operational problems, swing check valves are prohibited in some plants. A dual-plate check valve or an axial check valve could be considered as an alternative. Dual-plate check valves cause only low to moderate slamming, whereas axial check valves do not slam. In terms of capital cost or the cost of buying the valve, swing check valves are less expensive than dual-plate and axial flow check valves. It should be noted, however, that swing check valves are more likely to increase operational costs and hence the overall cost of the valve in comparison to the other two options. The most expensive check valves are axial valves.

3.4.2 Dual-Plate Check Valves

There are two spring-loaded disks in a dual-plate check valve, as illustrated in Figure 3.30. Can such a design significantly reduce water hammering? In the first instance, gravity does not close the valve disks. Rather, the disks are pushed back to their initial closed position by spring force. Second, dual-plate check valves

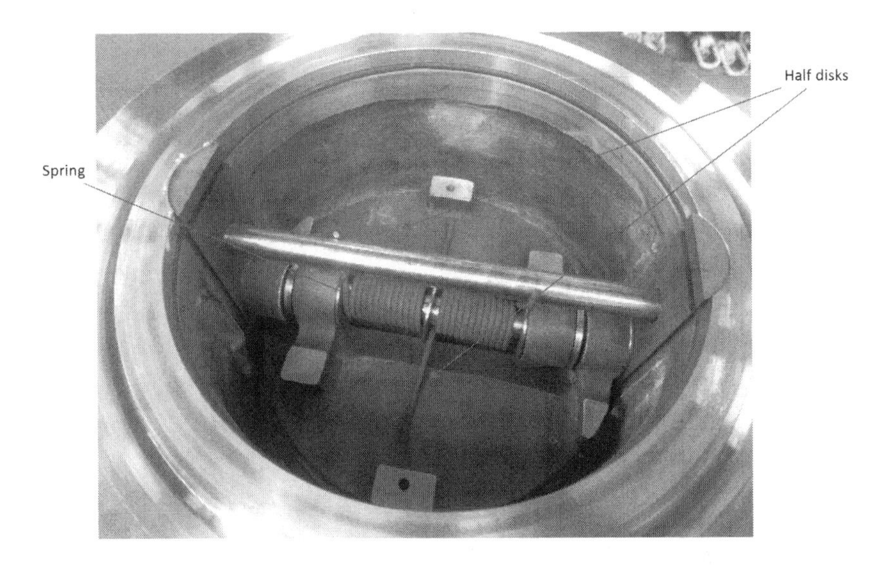

FIGURE 3.30 Dual-plate check valve with spring-loaded double disks.

have two disks instead of one, thereby distributing the weight of one disk over two disks, resulting in a less pronounced slamming sound. Fluid pressure is used to open dual-plate check valves. As soon as the fluid stops flowing, the spring force is coupled with the force of the reverse flow and closes the valve. In reality, the valve closes when the fluid rate decreases until the spring force overcomes the fluid pressure and prevents the plates from opening. Briefly, a dual-plate check valve is a non-return valve that is stronger, lighter in weight, and smaller in size in comparison to a conventional swing check valve. The double-plate check valve is composed of two spring-loaded plates that are hinged around a central pin. Backflow is prevented by the use of check valves after pumps and compressors. Pumps and compressors can be damaged by reversing flow from the discharge to the suction. It is not recommended to install dual-plate check valves after pumps and compressors, even though they can provide a low to moderate slamming effect. An axial check valve with a non-slamming characteristic is the most appropriate valve to install after pumps and compressors. The next section discusses axial check valves.

3.4.3 Axial or Nozzle Check Valves

Nozzle check valves are also known as axial flow check valves and are important for non-return fluid purposes with non-slamming and fast closing characteristics. In most cases, they are installed downstream of rotating equipment in order to prevent damage to the expensive mechanical equipment caused by backflow. Even though they are more expensive than other alternatives such as swing and dual-plate check valves, they can save a great deal by ensuring the safe operation of rotating equipment and reducing the pressure drop. In the oil and gas industry,

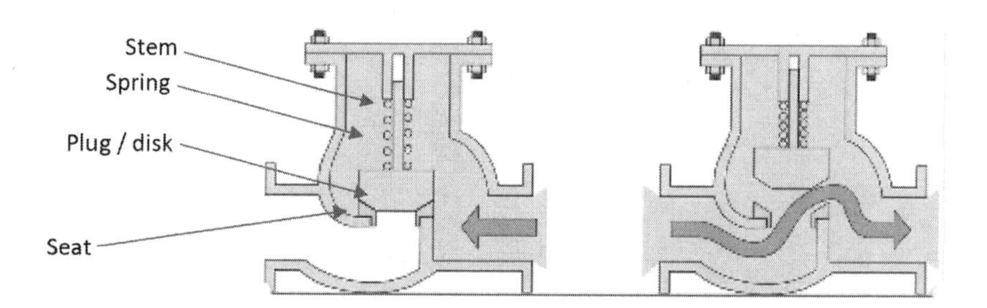

FIGURE 3.31 Non-slam check valve and the flow of fluid inside the valve.

these valves are widely used in a variety of sectors, including offshore platforms, subsea installations, refineries, pipelines, liquified natural gas (LNG) plants, and petrochemical plants.

The name of the valve is taken from its internal structure (see Figure 3.31). As illustrated in the figure, there is a nozzle inside the valve. Figure 3.31 illustrates how fluid flow comes from the right side and overcomes the spring force behind the disk to push the disk back and open the valve. During a flow stoppage or reverse flow, a spring pushes the disk back on its seat, and the valve closes.

The following are some of the main advantages of this valve:

1. ***Quick closing and opening:*** Short axial disk travel to the seat, spring-assisted design, and low mass disk make the nozzle check a fast-closing valve, which is particularly useful in critical lines or lines with fast reversing flows. A fast closing response reduces the likelihood of damage to equipment due to backflow and provides good protection for mechanical equipment that is expensive. As a result of the low static pressure behind the disk in the Venturi area, there is also a pressure differential over the disk, causing the disk to open easily.
2. ***Robust structure and zero leakage:*** The robust body structure of the valve provides greater resistance to vibration from upstream equipment (pumps and compressors) than dual-plate and swing check valves. Using an integrated body without a bolted body bonnet reduces the risk of leakage through the body to a very low level. These valves are used in subsea applications due to their zero-emission body characteristics.
3. ***Low pressure drop and high flow capacity:*** It is important to note that a high pressure drop causes the valves to wear and erode more rapidly as well as requiring the selection of more expensive pumps and compressors to provide higher head and lower friction loss. In normal circumstances, the valve has a pressure drop of less than 0.1 barg. Additionally, the Venturi effect associated with nozzle design (bore reduction) results in a smooth flow pattern through the valve, which eliminates flow turbulence, minimizes erosion problems, maximizes flow capacity, and facilitates easy opening. In summary, this valve is advantageous both from a process and mechanical perspective.

4. ***Long valve life:*** The maintenance requirements of non-slam check valves are low or even nonexistent from an operational perspective. Due to the only moving part being a disk that travels a very short distance to the seat, wear on the disk and the valve is kept to a minimum. Unlike dual-plate check valves, the spring is only exposed to axial load, increasing its lifespan and reducing wear. Some valves are returned to the manufacturers after a long period of operation, typically 7 to 10 years, for examination and possible maintenance. It is normal for these valves to be opened, examined, greased, and pressure tested and then sent back to the operator without requiring any major maintenance.
5. ***Non-slamming:*** In spite of the fact that the disk closing speed is fast, there is no slamming against the seat as a result of the short stroke and spring-assisted design, which minimizes water hammer and valve slam in the liquid services. Water or liquid hammer occurs when the fluid service produces a pressure wave as a result of sudden closing or slamming of the valve. There is a risk of noise, vibration, and/or piping collapse as a result of additional piping loads.
6. ***Shorter straight pipe run requirement:*** Ideally, check valves should be connected to straight pipes of 5 diameters upstream and downstream of the valves. A lack of straight pipe before the dual-plate and axial flow non-slam check valves, particularly when the fluid velocity is above 4.5 m/s, results in significant valve wear and a reduction in the design life. Since layout space is limited, achieving 5D of straight pipe is very challenging and sometimes not possible. The special geometry of the nozzle in the axial flow check valve, as discussed earlier, produces a Venturi effect and smoother flow characteristics, allowing just 2D or 3D straight pipes to be used upstream and downstream. Consequently, material, cost, and space are saved, all of which are critical parameters when it comes to offshore platforms.
7. ***Tight shutoff:*** Axial flow check valves are tight shut off, although they are metal seated.
8. ***Different end-to-end design (flexibility in end-to-end design):*** Despite the fact that valve manufacturers often choose a short pattern design (short pattern) for their products, ASME B16.10 (API 6D) (long pattern) end-to-end designs are also available on the market. When a compact nozzle check meets API594 requirements in end-to-end design, it may result in higher pressure loss and more internal wear than a manufacturer's end-to-end design. Compact designs are not recommended unless space is an important consideration. The pressure loss in accordance with ASME B16.10 or API 6D (long pattern) is approximately the same as the pressure loss in accordance with short pattern manufacturer standards. In the event of maintenance, it may be possible to order a new valve to replace the old one with a long pattern design from a different non-slam manufacturer, provided that the manufacturer end-to-end design is replaced by a valve from the same manufacturer.

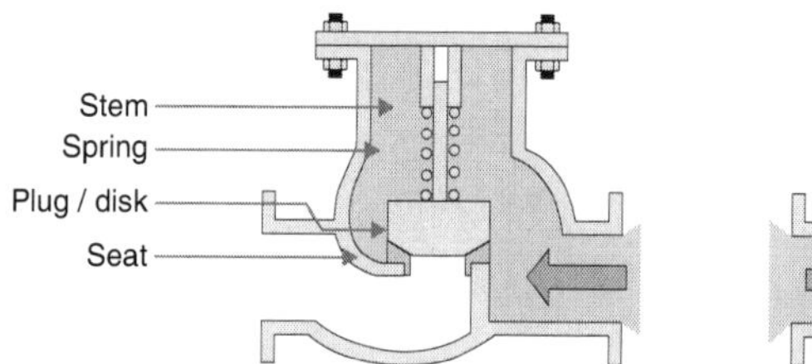

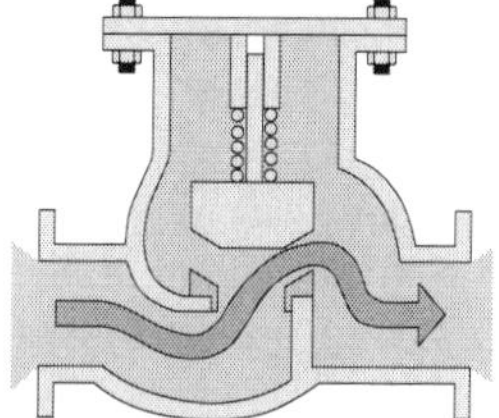

FIGURE 3.32 Piston check valve.

(Courtesy: Shutterstock)

3.4.4 Piston Check Valves

Small piping in sizes of 2" and less is typically not fitted with swing and dual-plate check valves, as they create considerable pressure drop inside the pipe. Instead, piston check valves are used for piping in small sizes. Figure 3.32 illustrates a piston check valve and its internal components: the stem, spring, disk or plug, and seat. The piston check valve opens when the fluid pressure and flow enter the valve from under the disk on the left side. Fluid pressure overcomes the spring force and pushes the disk and stem upward, which opens the valve and allows fluid to pass through the valve, as illustrated on the right-hand side of Figure 3.32. As the fluid pressure decreases, the disk moves down due to its weight and spring force, causing the valve to close as the disk sits on the seat. According to the figure on the left, the check valve remains closed when fluid enters from the right side, a process known as reverse flow.

3.5 PRESSURE SAFETY AND RELIEF VALVES

A safety valve is a valve that prevents overpressurizing of pressurized equipment such as pipes, pumps, compressors, turbines, and boilers. Pressure safety valves and pressure relief valves are collectively called safety valves. Chemical processing facilities cannot be immune from overpressure, which necessitates the provision of overpressure protection. The purpose of the pressure relief valve (PRV) installed on pressurized equipment (labeled Protected System in Figure 3.33) is to release overpressure gases or fluids from the equipment into the flare system. Whenever there is a problem with the pressure piping or equipment that leads to overpressure scenarios, the excessive pressure can damage or burst the expensive equipment and piping. In order to prevent undesirable events and enhance safety and reliability, PRVs and PSVs are installed over pipes and equipment. In the event that the pressure inside equipment or piping exceeds the allowable limit, the safety valve automatically opens and releases the excess pressure. According to API standard 520 part 1, the sizing of pressure relief valves is clearly defined. In accordance with API 520, it is essential to understand the similarities and differences between a relief valve and a safety valve. Both types of safety valves

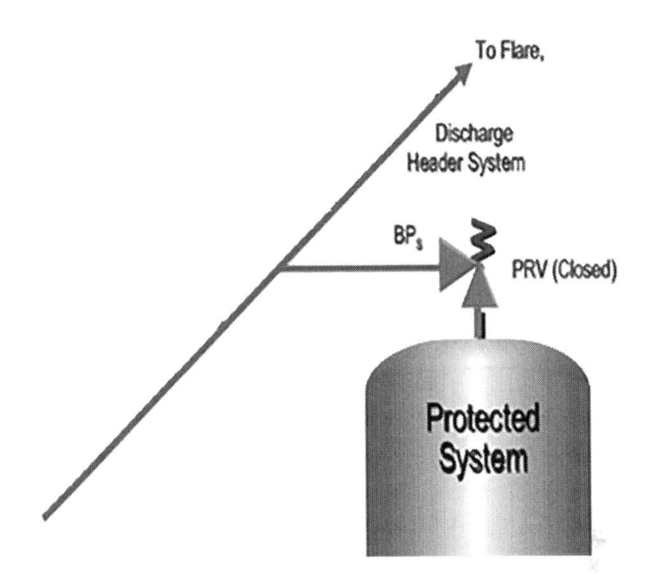

FIGURE 3.33 PRV installation on pressurized equipment connected to a flare system.

are spring-loaded and are designed to prevent overpressure in pressure piping and facilities. Relief valves are typically opened in proportion to the increase in pressure over the opening pressure. They are typically used for non-compressible fluids. The safety valve, on the other hand, is opened rapidly and is usually used for compressible services. It is possible to use a safety relief valve suitable for both applications. Gas flaring is used by operators to depressurize equipment and manage pressure variations that are unpredictable and large. In Chapter 5, you will find more information about gas flaring.

Three PSVs are installed on the piping system to protect it from overpressure in the event of an overpressure situation, as shown in Figure 3.34. Due to safety and reliability concerns, three PSVs are used instead of one. As a matter of fact, if one or two PSVs malfunction and require maintenance, the third one may be utilized. There are three ball valves located upstream of the PSVs that isolate the piping from the downstream PSV when it is removed for maintenance. A PSV is considered the last line of defense for overpressure prevention. This means that other overpressure safety devices, such as emergency shutdown valves, should prevent overpressure scenarios before any requirement for a safety valve to act.

A conventional spring-loaded pressure relief valve is illustrated in Figure 3.35. PSVs are similar to check valves in that they do not have a means of operation, such as an actuator, handwheel, or gearbox. Instead, PSVs operate automatically in response to changes in fluid pressure inside the connected piping. As illustrated in the figure, the valve has a body and a bonnet; these parts are pressure containing, which means that leakage from the body and bonnet leads to environmental pollution. There is a flow direction from the bottom of the valve under the disk or seat disk. When the fluid pressure exceeds the spring load located above the disk,

FIGURE 3.34 Three PSVs installed on piping for overpressure protection.

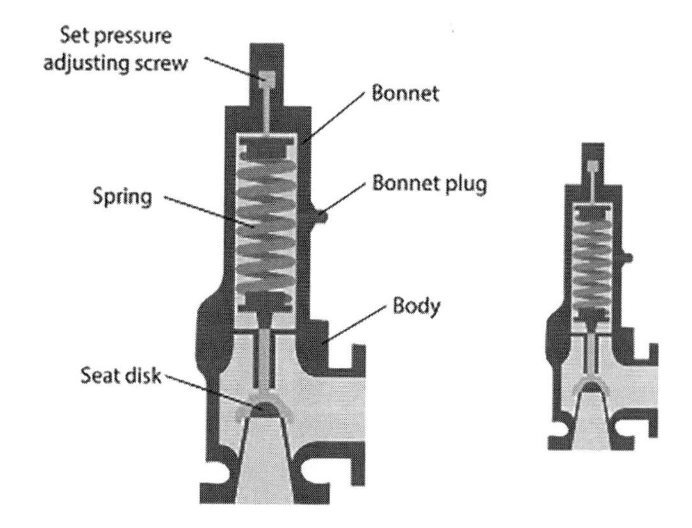

FIGURE 3.35 Conventional spring-loaded pressure relief valve.

(Courtesy: Shutterstock)

the disk is moved upward by the fluid pressure, and the fluid rotates 90° and exits the valve from the right side. In the event that the pressure inside the connected piping system drops back to a normal level, then the spring force will exceed the fluid pressure within the piping, pushing the disk downward, seated on the seat, and the valve will close. When the valve is closed, as shown in the figure, the disk sits on the seat area and prevents fluid from flowing.

As shown in Figure 3.36, a conventional safety valve includes the following essential parts: The main pressure-containment element is a body made of carbon steel or another corrosion-resistant alloy such as stainless steel. The pressure-containing components of valves are those components that, if they malfunction, could result in leakage from the valve into the environment. The fluid enters the valve through a port at the bottom, and it exits through the left port after rotating 90°.

Sizing pressure safety or relief valves requires consideration of both the valve size and the pipe size at the inlet and discharge. The inlet and outlet ports of a pressure relief or safety valve are typically different sizes, which results in the valve being identified as having two sizes, one for the inlet and one for the outlet. According to API 526, which deals with flanged steel pressure relief valves, pressure safety valve sizes are arranged alphabetically. According to ASME's Boiler and Pressure Vessel Code (BPVC), Sec. VIII, boiler construction rules are contained in the same letters as PSV sizing, that is, different orifice sizes for effective discharge. Based on the letters "D" through "T" in Table 3.1, API 526 orifice sizes are compared with ASME Sec. VIII orifice sizes. Despite having the same orifice letter, safety valves may have a variety of sizes of inlet and outlet connections. The orifice size of a 2" × J × 3" and a 3" × J × 4" safety valve is the same, as indicated by the letter J, but the size of the inlet and outlet are different, as shown before and after the orifice letter.

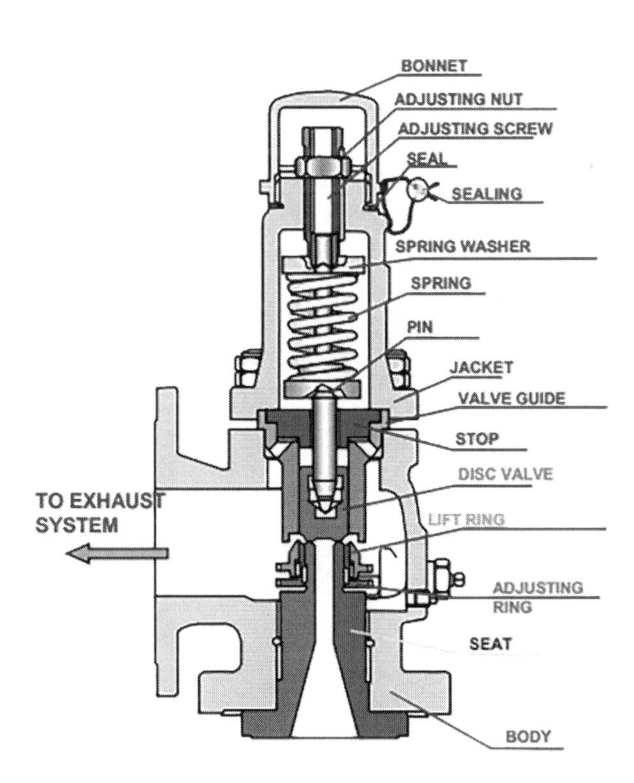

FIGURE 3.36 An essential component of a conventional safety valve.

TABLE 3.1
ASME and API Standard Orifice Sizes

Serial Number	Orifice Designation	API Effective Area (in²)	ASME Effective Area (in²)	PSV Inlet × Outlet Sizes (in)
1	D	0.110	0.124	1" × 2" 1.5" × 2" 1.5" × 2.5"
2	E	0.196	0.221	1" × 2" 1.5" × 2" 1.5" × 2.5"
3	F	0.307	0.347	1" × 2" 1.5" × 2" 1.5" × 2.5"
4	G	0.503	0.567	1.5" × 2.5" 1.5" × 3" 2" × 3"
5	H	0.785	0.887	1.5" × 3" 2" × 3"
6	J	1.287	1.453	2" × 3" 2.5" × 4" 3" × 4"
7	K	1.838	2.076	3" × 4"
8	L	2.853	3.221	3" × 4" 4" × 6"
9	M	3.600	4.065	4" × 6"
10	N	4.340	4.900	4" × 6"
11	P	6.380	7.205	4" × 6"
12	Q	11.05	12.47	6" × 8"
13	R	16.00	18.06	6" × 8" 6" × 10"
14	T	26.00	29.35	8" × 10"

QUESTIONS AND ANSWERS

1. Identify the correct statements regarding ball valve design considerations.
   A. Soft seats are more robust than metal seats.
   B. Top-entry ball valves can be accessed by removing the bonnet from the top after removing the bolts. In addition to being online maintenance friendly, this valve does not require removal from the line for maintenance.
   C. In sizes above 2", reduce bore ball valves are preferred. Reduced bore ball valves are lighter, more compact, and less expensive.

D. A floating design is more expensive than a trunnion design of the same size and pressure class.
E. A side-entry ball valve is characterized by the absence of a bonnet and can be constructed from two or three body pieces that are bolted together.
F. A fully welded body ball valve has a greater possibility of leakage than a side or top-entry ball valve.

Answer) Options B, C, and E are correct. As a result of the fact that metal seats are more robust than soft seats, option A is incorrect. The option D is incorrect, since a floating design is less expensive than a trunnion design. Due to the fact that fully welded valves have less leakage potential than both side-entry and top-entry valves, option F is incorrect.

2. Which parameter is not considered an advantage of an axial valve over a globe valve?
A. Less risk of cavitation.
B. Less possibility of leakage due to one-piece, bonnet-less design.
C. Shorter face-to-face/end-to-end dimensions.
D. More smooth flow inside the valve.

Answer) Options A, B, and D are all correct. An axial valve does not necessarily have a shorter face-to-face than a globe valve, making option C incorrect.

3. Which option gives all the essential characteristics of an axial check valve?
A. Non-slamming, compact, light, slow opening and closing
B. Smooth flow, high flow capacity, fast opening and closing
C. Non-slamming, fast opening and closing, robust design, low-mass disk
D. Medium slamming, short distance between the disk and seat, compact and light

Answer) It is not correct to select option A. While non-slamming is the main characteristic of an axial valve, axial valves are not necessarily compact and lightweight. Furthermore, axial check valves are capable of opening and closing rapidly. Option B is also not completely correct, as axial check valves do not have a high flow capacity. As a result of the valve internals, such as the disk and flow nozzle, being located on the flow path, the flow capacity is reduced. Nevertheless, check valves also possess two other characteristics, which are smooth flow and fast opening and closing. It is correct to choose option C, as it addresses all the essential characteristics of check valves. Axial check valves are not characterized by medium slamming and compact and light designs. Thus, option D is incorrect. Therefore, option C is the correct answer.

4. What type of valve is recommended for fluid isolation in a clean oil service at a pressure of 60 bars and a temperature of 150°C? Pipes connected to this valve are 20" in diameter and horizontally oriented. It is necessary for the valve's height to be as short as possible to avoid clashes with the pipe, as requested by the piping designer.
   A. Gate valve
   B. Ball valve
   C. Globe valve
   D. Plug valve

   Answer) Option A is not the best option because gate valves (expanding, slab, and wedge types) have a high height, which is a concern in this instance. It is incorrect to choose option C, since globe valves are used for flow regulation and not for flow isolation. Also, option D is incorrect since it proposes the use of a plug valve for particle-containing services, whereas in this instance, the fluid is clean. Therefore, option B, a ball valve, is the correct option.

5. Which sentences are correct about check valves?
   A. For valves smaller than 2", a dual-plate check valve is a suitable choice.
   B. As compared to dual-plate and swing check valves, an axial check valve can result in less pressure loss.
   C. Compared to dual-plate and axial check valves, swing check valves are the least expensive in terms of their initial or capital costs.
   D. By reducing the fluid pressure upstream of a dual-plate check valve, the valve will close as a result of the weight of the plates.

   Answer) It is not correct to choose option A. Dual-plate check valves can cause significant pressure drops in sizes of 2" and under, so they are not recommended in smaller sizes. Piston check valves are suitable for smaller sizes as low as 2" and under. Due to the Venturi effect inside axial valves, option B is completely correct, since axial check valves have the least pressure loss when compared to swing and dual-plate check valves. A swing check valve has a lower initial cost than a dual-plate or axial valve. Option C is also correct. There is a partial truth to option D, since when fluid pressure is reduced upstream of a dual-plate check valve, the valve is shut by spring force rather than by the weight of the plates. Therefore, options B and C are correct.

6. The oil and gas industry relies heavily on needle valves for flow regulation. What are the advantages of needle valves in this sector?
   A. Cavitation is not an issue with needle valves as it is with globe valves.
   B. The needle valve is capable of very precise regulation due to the taper-shaped plug or closure member.

C. It is possible to install these valves as a form of double block and bleed in order to save space, weight, and cost.
D. It is not necessary to apply high torque or force in order to operate these valves.

Answer) Since needle valves, like globe valves, are susceptible to cavitation, option A is not appropriate. Option B is the most appropriate choice, since needle valves are capable of very precise regulation due to the taper shape of the plug. The needle valve is used as a type of double block and bleed on a monoflange with the advantage of being lightweight, space saving, and cost effective. It should be noted, however, that having DBB needle valves on monoflanges does not add any value to the valve's ability to regulate flow. Despite the fact that needle valves do not require high torque or force to operate, this feature does not improve the valve's ability to regulate flow. Therefore, option B is the correct answer.

7. Which parameters make ball valves the better choice compared to gate valves?
A. More compact end-to-end length and actuator
B. More compact in height, less stem-sealing friction
C. Possibility of having either side-entry and top-entry design
D. Lower cost

Answer) It is incorrect to select option A, since ball valves do not necessarily have a shorter length than gate valves. In this case, option B is the correct option, and all the points outlined in this option are advantages of using a ball valve over a gate valve. While option C is a correct statement, it does not provide any advantages for ball valves over gate valves, which are always top-entry valves. Ball valves can be cheaper than gate valves of the same size and pressure class in many cases, but this is not true in every instance.

8. In this case, the main concern is to choose a compact valve for a seawater piping system with a pressure class of 150 equivalent to 20 bars. It is also important to consider the cost of the valve, which means that it should not be very costly. It is imperative that the required valve have an emergency shutdown safety-critical function and be capable of closing rapidly in the event of an emergency in order to shut down the line. For such an application, what type of valve would you recommend?
A. Ball valve
B. Concentric wafer butterfly valve
C. Double-offset wafer butterfly valve
D. Triple-offset wafer butterfly valve

Answer) Compared to butterfly valves, ball valves are more expensive, bulkier, and require more space. In this example, cost and compactness are critical parameters, which means that the use of a ball valve over a butterfly valve would increase both costs and space requirements. Considering that the fluid service is water (a non-process service), butterfly valves may be a good choice. With a butterfly valve, the flange can be eliminated, which is called a wafer design, which saves both space and money. Therefore, option A is incorrect. As a result of their rubber lining, concentric butterfly valves are not reliable valves and should be replaced in the event they become damaged. As a consequence, rubber-lined butterfly valves cannot be selected for safety-critical functions, so option B is not appropriate. There are both double-offset and triple-offset butterfly valves that are suitable for this application. However, triple-offset butterfly valves are more expensive than double-offset, and since cost is a concern here, a double-offset wafer butterfly valve is the most appropriate option. Therefore, option C is the correct answer.

9. Find the wrong statement about safety valves.
   A. When it comes to preventing overpressure, PSVs are considered the first line of defense.
   B. Like check valves, PSVs do not have a means of operation, such as an actuator, handwheel, or gearbox. Instead, PSVs operate automatically in response to changes in fluid pressure inside the connected piping.
   C. An overpressure scenario refers to a condition that would cause a pressure increase in the piping or pressure equipment beyond the specific design pressure or maximum allowable working pressure.
   D. PSVs release overpressure gasses from equipment in order to avoid overpressurizing and potential process safety incidents and to protect human life, property and the environment.

Answer) Option A is wrong because PSVs are considered the last line of defense.

10. Which of the following statements about cavitation is incorrect?
    A. Choke valves are highly susceptible to operational problems such as erosion and cavitation as a result of the 90° rotation of the fluid inside them.
    B. Axial valves have a low pressure drop, which prevents cavitation and loss of flow.
    C. API released the first edition of its new globe valve design standard (API 623) in 2013 to prevent and control problems associated with globe valve operation, including cavitation, vibration, and leakage.
    D. There are several methods of reducing cavitation, including reducing the diameter of the stem, strengthening the connection between the stem and plug, and applying hard-faced materials to the seat and plug, such as Stellite 6 or 21.

Answer) Option D is not entirely correct, since it is necessary to increase the stem diameter in order to avoid cavitation in globe valves.

FURTHER READING

1. American Petroleum Institute (API) 520. (2020). *Sizing, selection, and installation of pressure-relieving devices part 1—sizing and selection*. Washington, DC: API.
2. American Petroleum Institute (API) 521. (2007). *Pressure-relieving and depressuring systems*. 5th ed. Washington, DC: API.
3. American Petroleum Institute (API) 526. (2017). *Flanged steel pressure-relief valves*. 7th ed. Washington, DC: API.
4. American Petroleum Institute (API) 602. (2016). *Gate, globe and check valves for sizes DN100 (NPS 4) and smaller for the petroleum and natural gas industries*. 10th ed. Washington, DC: API.
5. American Petroleum Institute (API) RP 615. (2016). *Valve selection guide*. 2nd ed. Washington, DC: API.
6. American Petroleum Institute 609. (2004). *Butterfly valves: Double flanged lug and wafer*. 6th ed. Washington, DC: API.
7. American Petroleum Institute (API) 623. (2013). *Steel globe valves—flanged and buttwelding ends, bolted bonnets*. 1st ed. Washington, DC: API.
8. American Society of Mechanical Engineers (ASME) B16.34. (2020). *Valves—flanged, threaded, and welding end*. New York, NY: ASME.
9. American Society of Mechanical Engineers (ASME) B36.10/19. (2004). *Carbon, alloy and stainless-steel pipes*. New York, NY: ASME.
10. American Society of Mechanical Engineers (ASME). (2012). *Design and fabrication of pressure vessels: Boiler and pressure vessel code. ASME section VIII Div.02*. New York, NY: ASME.
11. Ballun, J. V. (2007). A methodology for predicting check valve slam. *Journal of American Water Works Association (AWWA)*, 99(3).
12. Crosby Valve Inc. (1997). *Crosby pressure relief valve engineering handbook*. Technical Document No. TP-V300. Wrentham, MA: Crosby Valve Inc.
13. Ford, R. (2014). Power industry applications: A valve selection overview. *Valve World Magazine*, 19(8), 96–103.
14. Hellemans, M. (2009). *The safety relief valve handbook: Design and use of process safety valves to ASME and international codes and standards*. 1st ed. Oxford, UK: Butterworth-Heinemann, and Imprint of Elsevier.
15. Nesbitt, B. (2007). *Handbook of valves and actuators: Valves manual international*. 1st ed. Oxford, UK: Elsevier.
16. Norwegian Oil Industry Association. (2013). *Valve technology*. 2nd revision. Stavanger, Norway: Norsk Olje & Gass.
17. Oxler, G. (2009). Non-return valve and/or check valve for pump system—a new approach. *Valve World Magazine*, 14(4), 75–77.
18. Provoost, G. A. (1982). *The dynamic characteristic of non-return valves*. Conference paper submitted to 11th symposium of the section of hydraulic machinery, equipment and cavitation, Amsterdam, The Netherlands.
19. Skousen, P. L. (2011). *Valve handbook*. 3rd ed. New York, NY: McGraw-Hill Education.
20. Smith, P., & Zappe, R. W. (2004). *Valve selection handbook*. 5th ed. New York, NY: Elsevier.
21. Sotoodeh, K. (2015). Axial flow nozzle check valves for pumps and compressors protection. *Valve World Magazine*, 20(1), 84–87.
22. Sotoodeh, K. (2017). Butterfly valves application in the Norwegian offshore industry. *Valve World Magazine*, 22(2), 63–64.
23. Sotoodeh, K. (2018). Selecting a butterfly valve instead of a globe valve for fluid control in a utility service in the offshore industry. *American Journal of Mechanical Engineering*, 6(1), 27–31.

24. Sotoodeh, K. (2018). Comparing dual plate and swing check valves and the importance of minimum flow for dual plate check valves. *American Journal of Industrial Engineering*, 5(1), 31–35. https://doi.org/10.12691/ajie-5-1-5
25. Sotoodeh, K. (2018). Why are butterfly valves a good alternative to ball valves for utility services in the offshore industry? *American Journal of Industrial Engineering*, 5(1), 36–40. https://doi.org/10.12691/ajie-5-1-6
26. Sotoodeh, K. (2019). Actuator sizing and selection. *Springer Nature Applied Science*, 1, 1207, Springer. https://doi.org/10.1007/s42452-019-1248-z
27. Sotoodeh, K. (2021). *A practical guide to piping and valves for the oil and gas industry*. 1st ed. Oxford, UK: Elsevier (Gulf Professional Publishing). ISBN: 978-0-12-823796-0
28. Sotoodeh, K. (2021). *Safety and reliability improvements of valves and actuators for the offshore oil and gas industry through optimized design*. PhD thesis, UiS No. 573, University of Stavanger, Stavanger.
29. Sotoodeh, K. (2022). *Coating application for piping, valves and actuators in offshore oil and gas industry*. 1st ed. Boca Raton: CRC Press (Taylor and Francis). ISBN: 9781032187198

4 Actuators

4.1 INTRODUCTION

There must be a means of operating every valve. All valves except pressure relief and check valves can be categorized as either manual or actuator operated. A manually operated valve is one that is operated by a lever or a handwheel connected to a gear box. An operator (human) is demonstrating a manually operated valve with a handwheel and a gearbox in Figure 4.1. The process of an actuator is sometimes compared to the functioning of a human body in order to answer the question of what an actuator does. Actuators perform mechanical actions in a machine in a similar manner to muscles in the human body that convert energy into motion.

Increasing efficiency and productivity in the oil and gas industry require more valve automation with actuators than in the past. Approximately 75% of oil and gas valves are automated by actuators, according to a report published by European Industrial Forecasting in 2015. Basically, a valve actuator is a black box with a signal receiving unit (air or oil) that produces some torque values for valve movement as an output. The quality of a valve is determined by many factors, including its metallurgy, mechanical strength, and machining, while its performance is largely determined by its actuator. As shown in Figure 4.2, an actuator is a mechanical or electrical component mounted on top of a valve for the purpose of automatically moving and controlling the valve's opening and closing. Therefore, actuators play an important role in valve automation. Valves can be automated using actuators in such a way that no human interaction is required for their operation. Today, actuators are used more widely than in the past, for a variety of reasons, such as the reduction of personnel requirements to operate valves; the ability to control valve operations precisely; and ease of operation, safety, and speed, as well as the reliability of the valves. For example, valves that require frequent operation or those located in remote or hazardous locations should be actuated. There are three principal factors that engineers must consider when selecting an actuator: frequency of operation, ease of access, and critical functions. A valve with a high number of operations is a good candidate for actuation. A valve actuator performs several functions, including moving the valve closure member to a suitable open or closed position, holding the valve closure member in the desired position, and providing sufficient force or torque to shut down the valve within an acceptable leakage class.

An actuator works by converting an external energy source into a mechanical motion, such as air, hydraulic power, or electricity. Valve quality is largely determined by the metallurgy and material selection of the valve components and on the mechanical strength of the parts, particularly pressure-containing parts, as well as by proper machining and tight tolerances. The actuators are responsible

DOI: 10.1201/9781003465881-4

FIGURE 4.1 A manual valve in the plant under operation by an operator.

FIGURE 4.2 Actuated valves.

for the performance of the valve, such as its ability to open and close, also known as cycling. Various types of actuators are found and used in processing plants, such as refineries, petrochemical plants, pipelines, and nuclear power plants. Various sources of power can be used to power actuators, including air, hydraulic oil, electricity, or a combination of hydraulic power and electricity, as explained in the following section.

4.2 ACTUATOR SOURCES OF POWER

Actuators in these sectors use four main power sources to control the industrial valves: air, hydraulic oil, electricity, and gas, all of which are briefly discussed in this section. When moving the actuator, it is important to consider the combination of electrical and hydraulic power sources.

4.2.1 Air

One of the main sources of power for actuators is compressed and pressurized air with a pressure value of around 5 to 9 bar. Air-operated actuators are known as pneumatic actuators. It is possible to use pneumatic actuators for both linear and quarter turn (rotary) valves. Pneumatic actuators have the advantage of working with air, which is a safe and environmentally friendly fluid, in contrast to hydraulic oil. Additionally, pneumatic actuators offer the advantages of fast operation and the ability to keep the valve closed or open even when the air supply is lost, known as a failsafe feature. In pneumatic actuators, there is a more compact control panel or system than in hydraulic actuators (see Figure 4.3 for a comparison of pneumatic and hydraulic control panels). Control panel function is explained later in this chapter. Compared to electrical actuators, pneumatic actuators have a higher torque-to-weight ratio. As hydraulic actuators contain fewer moving parts, they are less likely to wear out and require less maintenance.

Pneumatic actuators have two disadvantages: a relatively high cost compared to electrical actuators and less precise actuation due to the compressibility of air during operation. In the case of large-size and high-pressure valves, pneumatic actuators are limited by their capacity to provide sufficient torque or force. For large-size and high-pressure classes, pneumatic actuators should be replaced with hydraulic actuators. When it comes to selecting actuators based on size and pressure class, there is no clear differentiation between pneumatic and hydraulic actuators. This author has observed that a 24" slab gate valve in pressure class 1500 is equivalent to 250 bar pressure as specified in ASME B16.34 for valves, which will be installed on a topside platform in the offshore oil and gas industry. This slab gate valve is equipped with a hydraulic actuator since a pneumatic actuator of the same size as the hydraulic actuator in the figure cannot produce enough force to operate the valve.

Air leaking from the actuator does not harm the environment but has other negative consequences, such as increasing the cost of operation and potentially

FIGURE 4.3 Control panel/system of pneumatic actuator (left side) vs hydraulic actuator (right side).

harming or killing the operator. It is costly to lose compressed air in a plant. Thousands of dollars can be wasted each year as a result of compressed air leaks. When compressed air is lost, the pressure is lost, and the performance of the connected product, such as an actuator, is compromised. An example would be when the air pressure supplied to the actuators is lost, resulting in insufficient force to operate the valve connected to the actuator, so the valve cannot be fully opened or closed. According to estimates, up to 30% of the air flow outputs from compressors in an oil and gas unit are lost as a result of air leaks. In spite of this, not all of the air loss or leakage is related to pneumatic actuators and related systems. The Occupational Safety and Health Administration (OSHA) also states that air pressures greater than 30 psi, or almost 2 bar, are dangerous to humans. The lungs may be damaged by inhaling compressed air, and the ears and brain may be damaged by compressed air entering the ear. It is not common for subsea oil and gas companies to use pneumatic actuators. This is due to the fact that many of the valves in subsea areas are high pressure and should be operated by hydraulic actuators.

4.2.2 Hydraulic Oil

As with pneumatic actuators, hydraulic actuators convert fluid pressure into mechanical movement. There is no difference in the working principle of a hydraulic actuator and that of a pneumatic actuator, except for the source of fluid power. For hydraulic actuators, pressurized hydraulic oil with a pressure of 160 to 200 bar is used instead of pneumatic air. In comparison with pneumatic actuators, hydraulic actuators can produce a much greater force or torque when operating a valve. As a matter of fact, hydraulic actuators are a more suitable choice for valves requiring a high amount of torque or force to operate. A greater amount of torque is typically required for large-size valves in high pressure classes. The type of valve, in addition to the size and pressure class, affects the amount of force or torque required to operate the valve. In comparison to a butterfly valve of the same size and pressure class, ball valves have heavier internal components, such as closure members. Consequently, the torque required to operate a ball valve is typically greater than that required to operate a butterfly valve of the same size and pressure class. An example of a high pressure class and relatively large gate valve with a hydraulic actuator is shown in Figure 4.4.

Hydraulic actuators have the following advantages: Since oil pressure is higher than air pressure, hydraulic actuators are more compact than pneumatic actuators. Due to the higher oil pressure in a hydraulic actuator compared to an air actuator, less oil is required to generate the same amount of force compared to air. Since hydraulic oil is not compressible, hydraulic actuators are more precise than pneumatic actuators. Large-size and high pressure class valves can be operated at high speeds using hydraulic actuators. Compared to pneumatic and electrical actuators, hydraulic actuators have a higher speed. As compared to pneumatic and electrical actuators, hydraulic actuators have the highest torque-to-weight ratio. Hydraulic actuators contain fewer moving parts, so they are exposed to less wear and are easier to repair.

FIGURE 4.4 A slab gate valve with hydraulic actuator.

A hydraulic actuator's disadvantages are summarized as follows: The high pressure of hydraulic fluid handling requires an extremely high level of safety and precaution. Unlike leaks of air, hydraulic oil leaks have negative effects on the environment. The control system of a hydraulic actuator is usually a large box, as illustrated in Figure 4.3, which requires a large amount of space. A large space is always a challenge in the offshore oil and gas industry.

4.2.3 Electricity

Valves are operated by electrical actuators that use electricity or electrical power. It is possible for the electrical actuator or motor to operate at different AC or DC voltages. As a driving force, three-phase AC motors are commonly used. As an alternative, single-phase AC or DC voltage motors are available. When a higher torque or force value is required for the operation of the valve, electrical actuators, also called motors, are provided with a gear box. For example, an electrical motor may be sufficient to operate a 4" butterfly valve operated at pressure class 150, but a motor with a gear box should be upgraded for a 30" butterfly valve operated at pressure class 150 to provide a higher torque.

In summary, electrical actuators have the following advantages: They are relatively inexpensive, easy to operate, and available in a variety of plants. In contrast to hydraulic and pneumatic actuators, electrical actuators are the safest actuators and have no adverse effects on the environment, unlike oil, and electricity does not exert any pressure that may harm a human, unlike oil and air in hydraulic and pneumatic actuators, respectively. There is usually a screen on site that shows the percentage of valve opening for electrical actuators. Unlike pneumatic and hydraulic actuators, electrical actuators do not require a control system or accessories. When compared to pneumatic and hydraulic actuators, electrical actuators are less expensive, more compact, and lighter in weight. Using a control room, it is possible to monitor the status of the electrical actuators and the connected valves from a distance. Process Field Bus (PROFIBUS) is commonly used to transfer data from the electrical actuators to the control room. PROFIBUS is a digital network that facilitates communication between sensors and control systems. The PROFIBUS system for electrical actuators is typically supplied by the manufacturers of the actuators. A summary of PROFIBUS is that it provides a means of communication and control between electrical actuators. The PROFIBUS design card is inserted into the electrical actuators that need to be controlled in order to provide a network and wireless protocol for high-speed data communication between the actuators and the control room. When the PROFIBUS card is inserted into the actuator, the card will be in contact with the actuator's electronics and will receive all commands related to the actuator's movement. As an alternative to pipework and tubework, which are used for directing and routing air and hydraulic oil to the actuators, cables are more convenient to route electricity to the actuators.

There are several disadvantages associated with electrical actuators, including the following: Except for a few configurations that are equipped with springs or hydraulic power, electrical actuators are incapable of providing a failsafe position. Additionally, electrical actuators can be used in different sectors of oil and gas, except for subsea applications, to achieve fail-as-is or fail-at-the-last-position. It is possible to maintain a failsafe position with subsea electrical actuators. Furthermore, electro–hydraulic actuators can also maintain a failsafe mode of operation. In comparison to pneumatic and hydraulic actuators, electrical actuators contain more sensitive components. Compared to pneumatic and hydraulic actuators, electrical actuators have more complex internal components. As a result of the advent of new electronic technology, electrical actuators are changing. However, the constant development of electrical actuators with new technologies can be considered an advantage. When used in hazardous environments such as explosive areas, electrical actuators require more documentation and certificates than hydraulic or pneumatic actuators. Furthermore, electrical components are more hazardous in the event of a fire than pneumatic and hydraulic components. Actuators powered by electricity cannot provide stroke speeds as fast as those powered by pneumatics or hydraulics.

4.2.4 Gas

In most cases, gas actuators are used in industries such as petroleum refinement, where a ready supply of natural gas is available. Strictly speaking, a gas actuator operates solely by the combustion of gas. The term is, however, often applied to gas over oil actuators. Using a high-pressure gas source, these devices compress hydraulic oil, which is then used to drive the actuator mechanism. Gas inside the actuators can cause fugitive emissions; therefore this source of power is not recommended due to environmental concerns. As the term implies, fugitive emissions are unintentional, undesirable emissions, leaks, or discharges of gases or vapor from pressure-containing equipment or facilities, as well as from components within the plant such as valves, piping flanges, pumps, storage tanks, and compressors. A fugitive emission is also known as a leak or leakage.

4.3 ACTUATOR TYPES

It is possible to divide the motion of the valve stem into linear and rotating motions. Actuators should be designed and selected in accordance with the type of valve stem motion. As a result, linear and rotary actuators are generally used. For linear stem moving valves such as gates and globes, linear actuators should be used. Rotary stem valves, such as ball valves and butterfly valves, require actuators that can provide rotary motion, such as scotch yokes (scotch and yoke) and rack and pinions.

4.3.1 Rotary Actuators

There are several types of rotary actuators, including piston-type rotary actuators, diaphragm-type rotary actuators, and electrical actuators. It is possible for rotary piston actuators to be scotch-and-yoke, rack-and-pinion, or helical splines. In the following section, we will discuss rotary diaphragm–type actuators.

4.3.1.1 Rotary Diaphragm Actuators

In the oil and gas industry, rotary diaphragm actuators are used in all sectors except subsea. On the diaphragm side, rotary diaphragm actuators have the same mechanism as linear diaphragm actuators. As illustrated in Figure 4.5, linear motion is initially produced in the actuator and is then converted to rotary motion by a piston-crank mechanism. A linear diaphragm actuator, for example, closes a valve by pushing the valve stem down as the air pressure, which is transferred to the diaphragm, overcomes the spring torque under the diaphragm. When air supply fails,

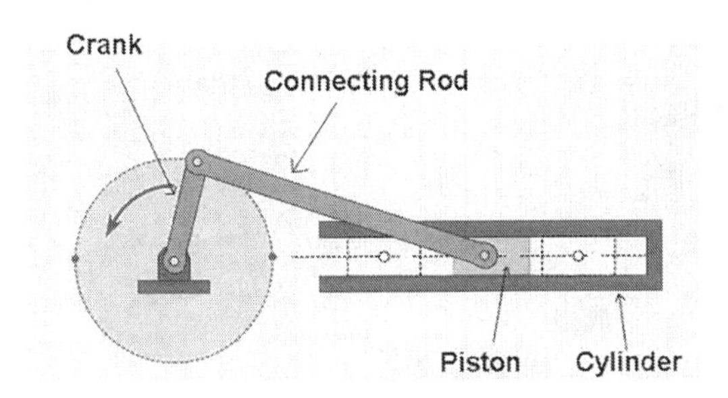

FIGURE 4.5 Piston crank mechanism.

the stem moves upward, causing the valve to open. The piston-crank system converts linear motion into rotary motion by connecting a piston with linear motion to a crank with rotary motion via a connecting rod, as illustrated in Figure 4.5.

4.3.1.2 Rotary Piston–Type Actuators

In this section, seven types of rotary piston actuators will be discussed: rack and pinion, scotch yoke, helical spline, electrical, electro–hydraulic, direct gas, and gas-over-oil. All of these actuator types are suitable for use in all oil and gas industries, including those in subsea, topside, refineries, and petrochemical plants.

4.3.1.2.1 Rack-and-Pinion Actuators

A rack-and-pinion actuator is a rotary actuator that can be powered by air or hydraulic oil. In accordance with its name, it has two gears (rack and pinion), as illustrated in Figure 4.6. Unlike the pinion, the rack has a linear gear movement that opposes the circular gear of the pinion. The linear movement of the rack is transferred through teeth to the gears in the pinion, such that rotary movement in the actuator and the connected valve is created in the pinion. It is possible for rack-and-pinion actuators to operate in either a failsafe or fail-as-is mode. In failsafe rack-and-pinion actuators, the rack is moved by air or hydraulic oil pressure on one side and a spring on the other side. A fail-as-is actuator has a double acting mechanism, which means that hydraulic oil is used from both sides to move the rack against the pinion. There is only one rack in the rack-and-pinion mechanism shown in the figure. As illustrated in Figure 4.7, a double or symmetrical rack can provide higher force and, more importantly, provide pressure balance.

FIGURE 4.6 Rack-and-pinion actuator.

(Courtesy: Shutterstock)

FIGURE 4.7 Rack and pinion with double rack.

(Courtesy: Shutterstock)

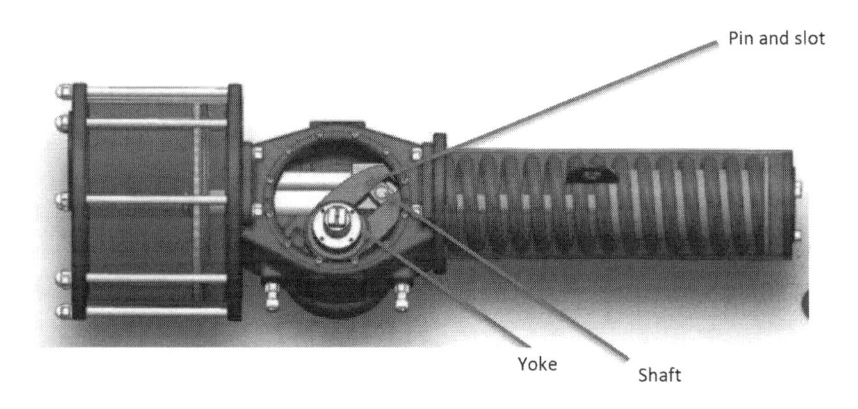

FIGURE 4.8 Scotch yoke spring return actuator.

4.3.1.2.2 Scotch and Yoke Actuators

This is another type of rotary actuator, suitable for quarter-turn valves such as ball and butterfly. This type of actuator can be used with hydraulic or air systems and may be either single acting or double acting. As shown in Figure 4.8, a spring-return scotch and yoke actuator converts linear or reciprocating motion into rotary or quarter-turn motion. An actuator's shaft is directly connected to its yoke through a slot or pin located in the center. By entering the actuator from the left side, air or hydraulic fluid moves the shaft forward. Through the slot or pin, the shaft's linear motion is transferred to the yoke. By rotating the yoke and connected shaft by 90°, the valve is moved between the open and closed positions.

Scotch and yoke actuators can be double acting (springless); that is, they can be pressurized from both sides by air or oil. Scotch and yoke actuators with a single-acting or failsafe mode of operation consist of a housing containing the yoke mechanism and a pressure cylinder containing the piston and spring.

4.3.1.2.3 Helical Spline Actuators

Helical spline actuators are a type of rotary valve that can be used for quarter-turn valves such as ball valves. The advantage of helical spline actuators over rack-and-pinion or scotch yoke actuators is that helical splines are mounted vertically over the valve, which is connected to a horizontal line. By using helical spline actuators, we are able to save a substantial amount of horizontal space that is consumed by rack-and-pinion actuators and scotch yoke actuators. Furthermore, helical spline actuators are lighter than two other types of rotary actuators. Helical spline actuators are, however, generally more expensive than rack-and-pinion actuators and scotch yoke actuators.

Typical helical spline actuators have internal gears that are connected to the external section of the spline, as illustrated in Figure 4.9. As hydraulic fluid enters from the top, it pushes the spline downward, and helical gearing causes the spline to rotate simultaneously with the shaft. The actuator is equipped with two helical gears: one between the external part of the spline and the internal part of the actuator body or casing, the other between the external section of the shaft and the internal section of the spline. A valve will be rotated by the rotation of the bottom of the shaft. An actuator with a helical spline can be installed vertically on a valve with a rotating motion, as shown in Figure 4.10.

4.3.1.2.4 Rotary Electrical Actuators

Several types of electrical actuators can provide rotary motion for quarter turn valves, such as ball valves. A valve actuator of this type can be installed on a

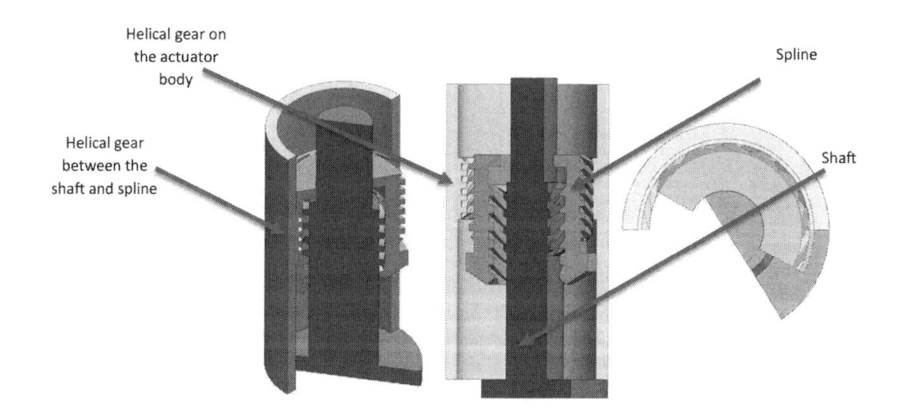

FIGURE 4.9 Helical rotary actuator internals.

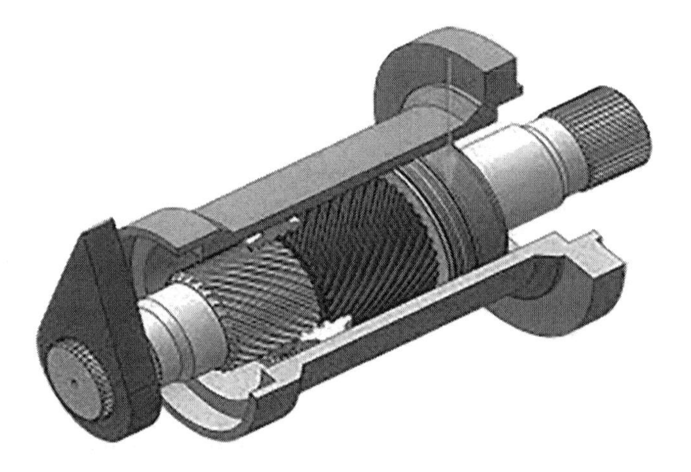

FIGURE 4.10 Helical spline actuator.

FIGURE 4.11 Electrical actuator on a large 38" ball valve.

variety of valve sizes, ranging from small valves of 1" to large valves of 38" or even larger. It is possible to adjust the output torque by adding more gears or by changing the gear ratio. AC or DC voltage can be used to operate the motors in electrical actuators. There are a variety of materials that can be used for the housing, including cast iron, carbon steel, and aluminum. A part-turn actuator attachment should be used for the flange connection between the actuator and the valve in accordance with ISO 5211. In Figure 4.11, an electrical actuator is used to automate a 38" ball valve. The purpose of an electrical actuator for such a large

FIGURE 4.12 Electro–hydraulic actuator.

valve is to facilitate operator operation. Manually opening and closing such a large valve would require a long period of time and many turns of the handwheel if it were supplied with a handwheel and gear box.

4.3.1.2.5 Electro–Hydraulic Actuators

An electro–hydraulic actuator is a type of hydraulic actuator that operates with electric motors; in fact, an electro–hydraulic actuator consists of two connected and integrated devices: an electric motor and a hydraulic actuator. An electro–hydraulic system has the main advantage of eliminating the need for hydraulic oil distribution systems, including pumps and tubing. The electrical motor can be used for providing power to move the hydraulic oil. Additionally, electro–hydraulic actuators are safer and more reliable than hydraulic actuators. In contrast to electrical actuators, electro–hydraulic actuators provide a failsafe mode of operation. In Figure 4.12, a hydraulic actuator and an electrical motor are integrated into an electro–hydraulic actuator.

4.3.1.2.6 Direct Gas Actuators

Actuators of this type operate by using high-pressure gas or nitrogen to open or close valves. Typically, the actuator is supplied with gas through a gas transmission system, such as a gas pipeline. In Figure 4.13, a rotary direct gas scotch yoke actuator is shown that is pressurized by oil from the left and gas from the right. Direct gas actuators are normally used for double-acting actuators, as illustrated in the figure. As a result of the use of gas inside a direct gas actuator, fugitive emissions can be generated. Direct gas actuators could be linear or rotary. Linear direct gas actuators are suitable for gate valves, while rotary direct gas actuators are suitable for quarter-turn valves like ball valves.

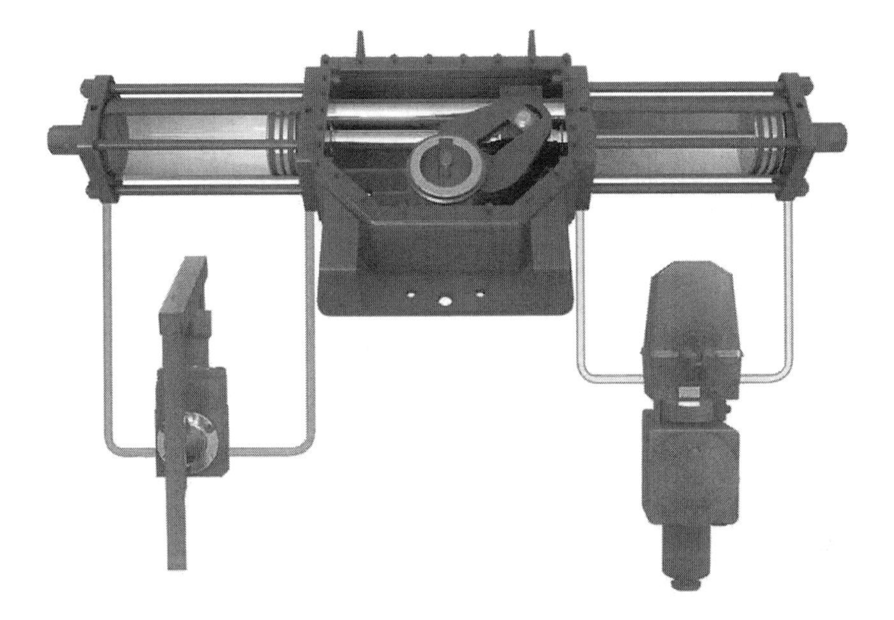

FIGURE 4.13 Direct gas actuator by Schlumberger.

4.3.1.2.7 Gas-over-Oil Actuators

The gas from the pipeline is used to pressurize the hydraulic fluid to move a double-acting hydraulic actuator. The use of high-pressure gas eliminates the need for pumps and additional tubing to pressurize hydraulic oil. A gas-over-oil actuator is illustrated in Figure 4.14 for a double-acting scotch yoke actuator. The use of gas inside a gas-over-oil actuator could be a source of fugitive emissions in the same manner as with rotary direct gas actuators. Natural gas pressure and rotary gas-over-oil actuators are used to operate valves in remote areas, such as those on pipelines.

4.3.2 Linear Actuators

This section discusses three types of linear actuators: linear pistons, linear diaphragms, and linear electrical actuators.

4.3.2.1 Linear Piston Actuators

A linear piston actuator can be powered by either air or hydraulic energy, meaning that they are either pneumatic or hydraulic in nature. The choice between hydraulics and pneumatics is dependent upon several factors, including the type of valve, its size, and its pressure class. The hydraulic fluid can produce a higher force when used to operate actuators and valves than pneumatic power. Therefore, a hydraulic actuator may be a more suitable choice for a type of valve such as a ball valve, which requires additional force to operate, especially in large sizes

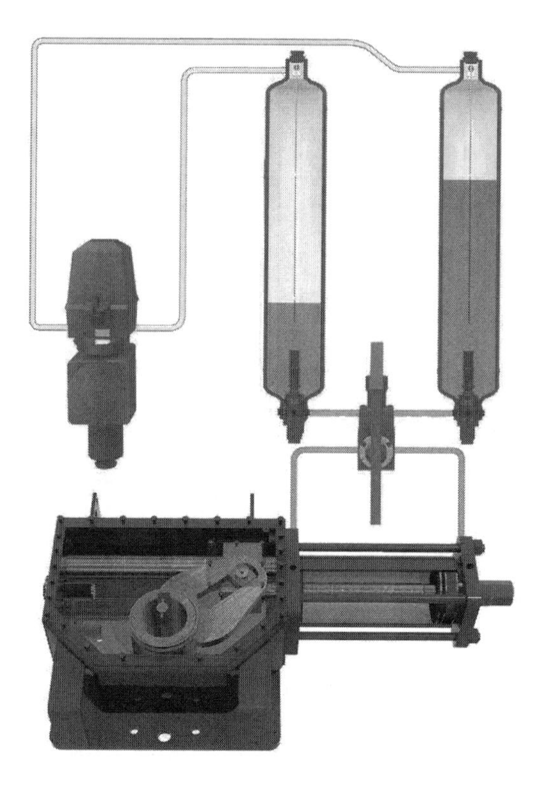

FIGURE 4.14 Gas-over-oil actuator by Schlumberger.

and high pressure classes. A linear piston actuator creates motion in a straight line and is very common for gate valves. It is common for linear piston actuators to be installed vertically on valves. They may be either single-acting, also known as spring return actuators, or double-acting actuators. As explained before, single-acting actuators are pressurized by fluid on one side and returned to fail-safe mode by spring force on the other. As opposed to single-acting actuators, double-acting actuators operate without springs and are pressurized from both sides by the fluid. The springless design of double-acting actuators prevents a failsafe mode in the event of a failure of the supply fluid. The failure mode of double-acting actuators is failure-as-is or failure-staying-in-position. A spring return hydraulic actuator with a linear stem movement is illustrated in Figure 4.15. Despite the fact that the actuator in the figure is mounted horizontally, the actuator is typically mounted vertically on a valve. Hydraulic oil enters the chamber on the left and overcomes the spring force, resulting in a movement of the piston rod to the left, which is transmitted to the valve by the movement of the stem rod. As soon as the flow of hydraulic fluid ceases, the spring force will overcome the hydraulic force and open the valve in the case of fail open actuators and close it in the case of failsafe close actuators.

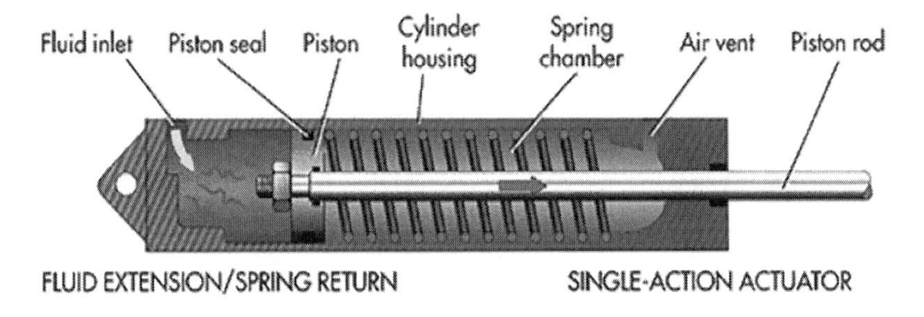

FIGURE 4.15 Linear spring return actuator.

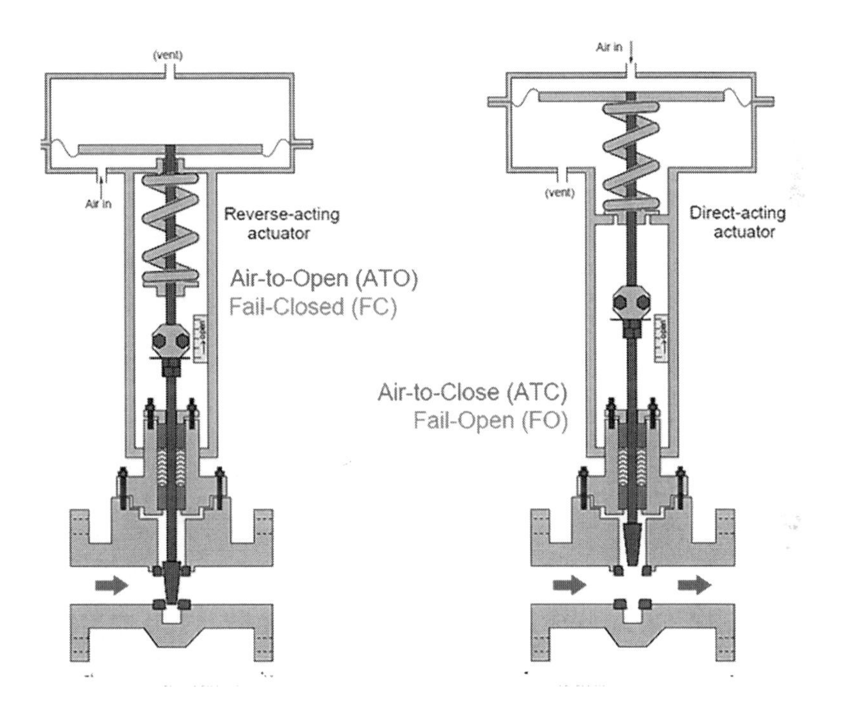

FIGURE 4.16 Comparing direct-acting and reverse-acting diaphragm actuators.

4.3.2.2 Linear Diaphragm Actuators

A diaphragm actuator is operated pneumatically and is supplied with air from the control system or another source. A diaphragm actuator is typically used for control valves, which are globe valves that are typically used to adjust process variables, such as pressure, temperature, or flow rate. In the oil and gas industry, diaphragm actuators are used in all sectors except for subsea. The two most common types of diaphragm actuator for control valves are direct acting (right side) and reverse acting (left side). Figure 4.16 compares direct-acting (right side) and reverse-acting (left side) diaphragm actuators. In direct-acting actuators, the air enters the top area of the diaphragm and pushes it downward. As air pressure is transferred to the diaphragm, it overcomes the spring torque located under the diaphragm, which closes the valve stem. In the event that the air supply fails, the stem moves upward, and the valve opens. As a result, direct-acting diaphragm actuators are suitable for applications involving air-to-close (ATC) and fail-open (FO). The majority of control valve actuators are reverse acting, resulting in a fail-closed (FC) failure mode. The air supply port is located under the diaphragm in reverse port diaphragm actuators, which means that the air supply opens the valve, and the air supply is interrupted, resulting in the valve closing. Three main parameters determine the amount of force produced by this type of actuator: air pressure, diaphragm diameter, and spring force.

4.3.2.3 Linear Electrical Actuators

It is possible to use electrical actuators with liner stem movement on through conduit gate valves, as illustrated in Figure 4.17. In the motor, rotary motion is converted into linear motion. A linear electrical actuator's output thrust is determined by the gear ratio within the actuator. A flange connection is used between the electrical actuator and the valve (see Figures 4.17 and 4.18), according to ISO 5210, a multi-turn valve and actuator attachment. In the figure, valves and actuators are illustrated for topside and offshore applications in the oil and gas industry.

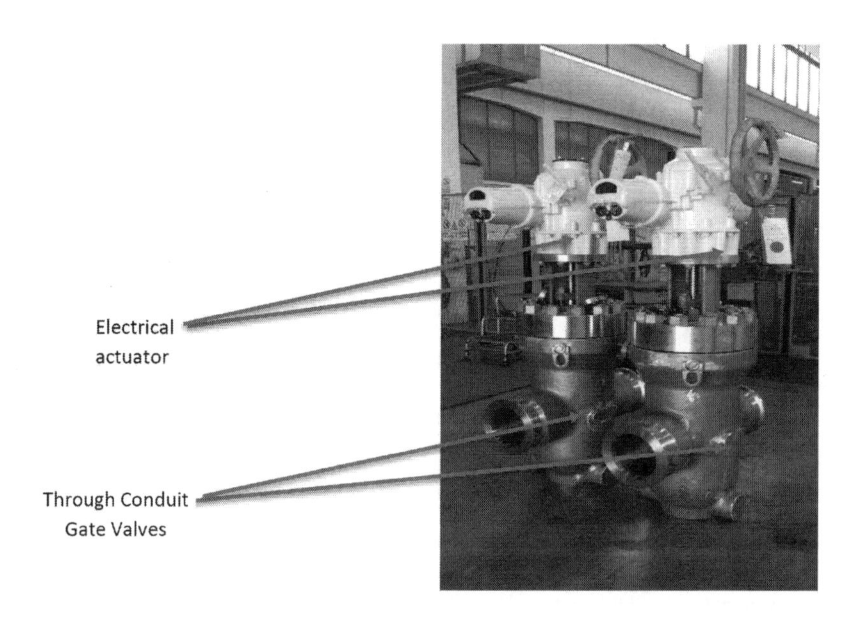

FIGURE 4.17 Linear electrical actuators on the through conduit gate valves.

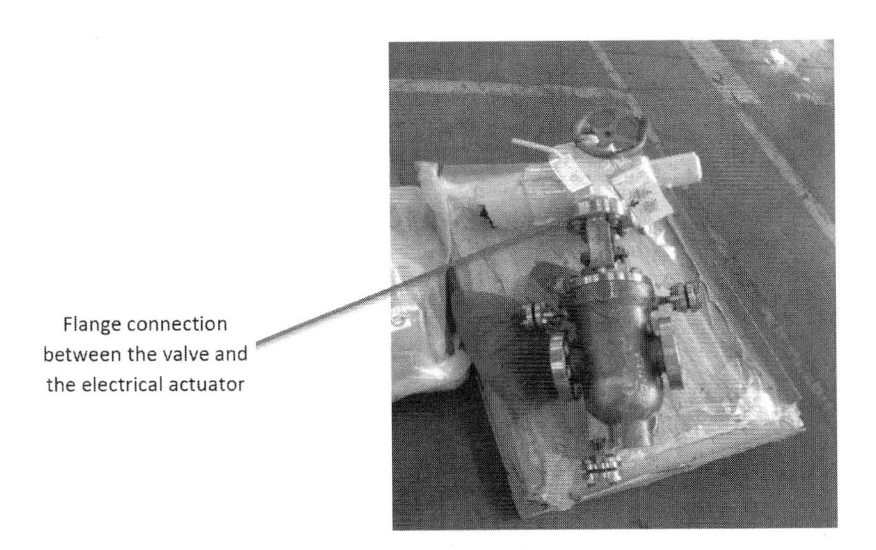

FIGURE 4.18 Flange connection between the valve and the electrical actuator per ISO 5210.

4.4 ACTUATOR ACCESSORIES (CONTROL SYSTEMS)

The purpose of this section is to explain the components that make up the control system or control panel. It is important to note that electrical actuators do not have a control panel as a separate unit, so the control systems discussed in this section apply only to pneumatic and hydraulic actuators. Unlike pneumatic and hydraulic actuators, electrical actuators come with integrated control accessories. Unlike mechanical actuators, electrical actuators do not require a separate control system, which is a significant advantage. Moreover, the control components discussed in this section can be applied to actuators in other sections of the oil and gas industry, except for subsea applications. An illustration of a pneumatic scotch yoke actuator for a butterfly valve is shown in Figure 4.19. As mentioned earlier, hydraulic actuators require a similar control system arrangement. However, hydraulic actuator control panels are larger than those for pneumatic actuators.

4.4.1 Block Valves

It is the job of block valves, which could be ball valves or needle valves, to move the source of energy (air or hydraulic) within the control panel in order to operate the actuator. As can be seen in Figure 4.19, the block valve is a ball valve that has been closed. When the ball valve is closed, it isolates the source of energy, allowing the actuator to return to either a fail-open or closed position in the event that the air supply is lost. However, the question is: How can the position of the ball valve be determined? An indication of the direction of the hole within the

FIGURE 4.19 Control panel and its components on a pneumatic actuator.

ball can be seen on the lever of the ball valve in blue. Because the lever of the ball valve is perpendicular to the flow of air in the control panel, the hole of the ball in the ball valve should be perpendicular to the flow of air in the control panel, which indicates that the ball valve is closed. It is common for the control panel to contain more than one valve in addition to the block valve. The check, solenoid, and quick exhaust valves (applicable only to pneumatic actuators) are also located on the control panel and will be explained later in this chapter.

4.4.2 Filter and Pressure Regulators

In both pneumatic and hydraulic actuators, this component removes dust and dirt, such as particles and rust, from the air or hydraulic oil. It is important to maintain a high level of cleanliness in both the hydraulic oil and the air, as particles in the hydraulic oil or air can interfere with the functionality of the actuators. Different standards may be used to define the cleanliness classification of hydraulic fluids, such as SAE 4059. A regulator is integrated with the filter to adjust the pressure of the supply fluid to the actuator. It is possible to see and measure the pressure of the supply fluid to the actuator, as shown in Figure 4.19.

4.4.3 Check Valves

The check valve is a non-return valve that operates only in one direction. The check valve opens when the supply fluid passes through it from the left side; however, the valve prevents the fluid from returning to the upstream side of the valve once it has passed through it. Figure 4.19 illustrates the installation of a check valve after the filter and pressure regulator.

4.4.4 Solenoid Valves

It is an electromechanical valve that is opened and closed by the start or stop of electricity. In pneumatic actuators as well as hydraulic actuators, solenoid valves are used.

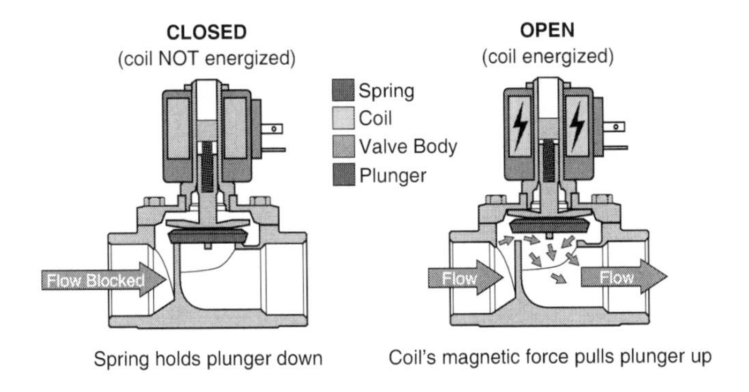

FIGURE 4.20 Solenoid valve functioning.

There is only one solenoid valve on the control panel, as shown in Figure 4.19; however, there may be two or three solenoid valves if higher safety features, such as an emergency shutdown feature, are required. An illustration of the functioning of a solenoid valve can be found in Figure 4.20. A solenoid valve is shown on the right under a condition of electricity being supplied to the coil, which energizes the coil and creates a magnetic field which surrounds the plunger. Through the magnetic force of the coil, the plunger is pulled up, and the valve is opened by movement of the plunger toward the top. When the magnetic field is eliminated by loss of electricity, the plunger moves downward, and the valve closes, as shown on the left side of the figure.

4.4.5 Quick Exhaust Valves

In order to achieve rapid piston speed on the return stroke of single-acting and double-acting pneumatic actuators, quick exhaust valves are used on the control panel to release the air more quickly. There are no quick exhaust valves on hydraulic actuator control panels due to the presence of hydraulic oil, which is not environmentally friendly.

4.4.6 Limit Switches

Due to the fact that a limit switch is not located on the control panel, it is not illustrated in Figure 4.19. According to Figure 4.21, limit switch boxes are sensors installed on pneumatic and hydraulic actuators to control their motion. The main function of a limit switch is to prevent the device from exceeding its mechanical movement range. For instance, a limit switch should not allow the actuator to move in a manner that would result in the ball of a ball valve connected to the actuator rotating more than 90°.

4.4.7 Tubes

Control panels are typically connected to actuators using a small tube with a maximum diameter of 1". There may be tubes connecting the components on

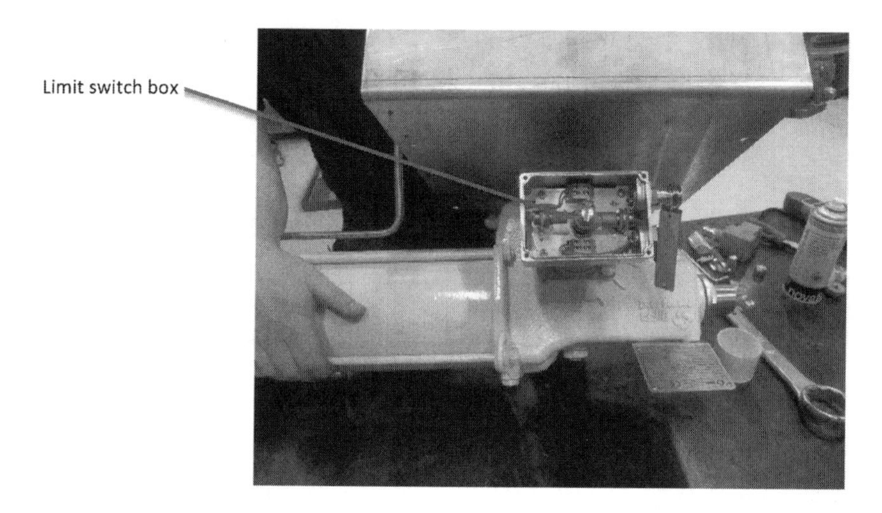

FIGURE 4.21 Limit switch on an actuator.

the control panel, or there may not be tubes connecting the components. There are no tube connections between the components on the control panel shown in Figure 4.19. The material of the tubing could be austenitic stainless steel grade 316 or 6Molebdeneym (6MO) alloy. Compared to stainless steel 316, 6MO is a super-austenitic alloy with a much higher corrosion resistance and mechanical strength. Some clamps are capable of supporting tubes at certain lengths. Offshore environments can result in stress cracking corrosion of tubes located under clamps, which is why the tubing material can be upgraded to 6MO from stainless steel 316 in some cases.

4.4.8 End Stoppers

There is no end stopper on a control panel. In order to completely adjust the valve closure member in the correct position, the end stopper is located at the end of the actuator cylinder. The actuator, for example, moves the ball of a ball valve 89° to the closed or open position. It is possible to adjust the 1° offset of the ball by rotating the end stopper on the actuator.

4.4.9 Position Indicators

Rather than being located on the control panel, the position indicator can be found on the actuator. It displays the position of the actuator. As an example, the position indicator shown in Figure 4.22 indicates that the actuator and connected valve are in the open position. On some of the valves that are actuated, proximity switches may be installed to report the position of the actuators to the control room.

FIGURE 4.22 Actuator position indicator showing open position on the top of the actuator.

4.5 ACTUATOR SELECTION PARAMETERS

Some may describe a valve actuator as simply a black box, consisting of an input (power supply or signal), an output (torque), and a mechanism to operate the valve. It will be evident to those who select valves that a variety of actuators are available to meet the needs of most individuals or plants for valve automation. The engineer must be aware of the factors that are most important when selecting actuators for plant-wide valve automation in order to make the best technical and economic choice. It is important to consider the following parameters when selecting an actuator for a valve:

1. ***Torque and size of the valve:*** In order to operate valves of large size and high pressure class (such as a 30" Class 1500 ball valve), a high torque or force is required. According to physics and mechanics, torque is the rotational equivalent of linear force. It is also known as the moment of force (abbreviated moment). In order to operate such a large valve, it is not economically feasible to select a very large pneumatic actuator. In this case, a hydraulic actuator is recommended.
2. ***Failure mode:*** In contrast to electrical actuators, pneumatic and hydraulic actuators remain in an open or closed position during a power outage. These valves are spring-return valves, which means that if there is a power failure or signal failure, the spring will return the valve to a predefined position. As a result, electrical actuators are not suitable for emergency shutdown valves that should be fully closed in the event of a power failure.
3. ***Cost:*** The cheapest actuators are electrical actuators, while the most expensive actuators are hydraulic actuators.

4. ***Availability of power source:*** In selecting an actuator, it is important to ensure that the actuator's power source is available. The use of a hydraulic actuator on a plant is not possible without a source of supply of hydraulic oil, for example.
5. ***Speed of operation:*** It is important to note that electrical actuators operate valves more slowly than pneumatic or hydraulic actuators, so an electrical actuator may not be an appropriate option if the valve is expected to operate at a speed of 1" per second.
6. ***Frequency and ease of operation:*** Many large-size valves that are operated frequently are operated with electrical actuators instead of manually. An electrical actuator is proposed to be mounted on a 20" class 300 manual ball valve for ease of operation, for example.
7. ***Control accessories:*** The control accessories of electrical actuators are integrated into the actuator, unlike those of pneumatic or hydraulic actuators. An advantage of electrical actuators is that they do not require any space for control accessories. There is a larger control panel on hydraulic actuators compared to pneumatic actuators.
8. ***Hazardous areas:*** There may be some limitations to the use of electrical actuators in hazardous environments.

QUESTIONS AND ANSWERS

1. Fill in the gaps with the correct words.
   A 38" ball valve in a high pressure class of 2500 with a failsafe closed function requires a/an ____ actuator for operation. This type of actuator is categorized as _____. The other type of actuator, which provides a fail-as-is function, is _______. This type of actuator with fail-as-is function can provide both _______ and linear motions.
   A. Electrical, linear, pneumatic, rotary
   B. Hydraulic, rotary, electrical, rotary
   C. Pneumatic, rotary, hydraulic, linear
   D. Electrical, linear, electrical, linear

   Answer) It is generally not possible for electrical actuators to provide a failsafe closed mode of operation. When the 38" ball valve has been actuated and is failsafe closed, pneumatic or hydraulic actuation options remain. As a 38" ball valve in pressure class 2500 is a large valve in a high pressure class, it requires a significant amount of force to operate; pneumatic actuators are unable to generate the force necessary to operate such a large valve at a high pressure level. Therefore, hydraulic actuators are the correct choice. Ball valves are quarter-turn valves with rotary action. A second type of actuator that provides fail-as-is functionality is the electrical actuator. Because electrical actuators are capable of both linear and rotary motion, option B is the correct answer.

2. Which types of actuators are not environmentally friendly?
   A. Hydraulic actuators
   B. Electro–hydraulic actuators
   C. Pneumatic and electrical actuators
   D. A and B are correct

   Answer) There are several environmental concerns associated with hydraulic actuators. Hydraulic oil spills can be absorbed by the ground and contaminate the water and food supply. In addition, electro–hydraulic actuators are not considered environmentally friendly; they are a type of hydraulic actuator that uses electric motors instead of pumps to move hydraulic oil. Both pneumatic and electrical actuators are environmentally friendly. Therefore, option D is the correct answer.

3. In relation to electro–hydraulic actuators, which of the following statements is true?
   A. The electro–hydraulic actuator is an electrical actuator that is powered by electricity.
   B. As a result of a simpler system architecture, electro–hydraulic actuators are less reliable than hydraulic actuators.
   C. The use of electro–hydraulic actuators eliminates the need for a separate pump and additional tubing.
   D. Electro–hydraulic actuators are a type of hydraulic actuator that work with hydraulic fluid.

   Answer) Since electro–hydraulic actuators are not electrical actuators, option A is incorrect. It is not correct to choose option B, since electro-hydraulic actuators have a higher level of reliability than hydraulic actuators. It is correct to choose option C since electro–hydraulic actuators eliminate the need for a separate pump and additional tubing. Considering that electro–hydraulic actuators are a type of hydraulic actuator that uses hydraulic fluid, Option D is also correct.

4. In relation to hydraulic actuators, which of the following statements is incorrect?
   A. Due to the higher oil pressure than air pressure, hydraulic actuators are more compact than pneumatic actuators.
   B. The higher oil pressure in a hydraulic actuator compared to an air actuator results in a smaller amount of oil being required to generate the same amount of force.
   C. Due to the fact that hydraulic oil is not compressible, hydraulic actuators are more precise than pneumatic actuators.
   D. All three choices are correct.

   Answer) Option D is the correct answer.

5. In relation to electrical actuators, which of the following statements is true?
   A. There is no failsafe position provided by electrical actuators, with the exception of a few specific configurations that are equipped with springs or hydraulic power.
   B. Electrical actuators contain less sensitive components compared to pneumatic and hydraulic actuators.
   C. Electrical actuators require less documentation and certificates in comparison to both hydraulic and pneumatic actuators, when they are used in hazardous environments such as explosive areas.
   D. Electrical actuators cannot provide stroke speed as fast as pneumatic and hydraulic actuators.

   Answer) It is correct to choose option A. It is incorrect to select option B since electrical actuators contain more sensitive components than pneumatic or hydraulic actuators. It is also incorrect to choose option C, since electrical actuators are required to provide more documentation in explosive areas than hydraulic and pneumatic actuators. It is also correct to select option D. Therefore, both options A and D are valid.

6. Which components are not located on a pneumatic actuator control panel?
   A. Limit switch, tubing, quick exhaust
   B. Limit switch, position indicator
   C. Pressure gauge, filter and air regulator
   D. Solenoid and block valve

   Answer) The pneumatic actuator control panel contains tubing, quick exhaust, pressure gauge, filter and air regulator, solenoid, and block valves. Due to the fact that the limit switch and position indicator are located outside the control panel, option B is the correct answer.

7. Identify the correct statement regarding rack-and-pinion actuators.
   A. The rack-and-pinion actuator is a linear actuator composed of two types of gears.
   B. One of the gears is the rack, which has a circular gear movement that moves against the pinion.
   C. Rack-and-pinion actuators work with hydraulic, air, and electricity.
   D. Rack-and-pinion actuators could be single or double acting.

   Answer) Option A is incorrect since the rack-and-pinion actuator is a rotary actuator and not a linear actuator. In addition, option B is also incorrect, since a rack has a liner movement and not a circular movement. Due to the fact that rack-and-pinion actuators are not powered by electricity, option C is not correct. The correct answer is option D.

8. In Figure 4.23, two types of actuators are illustrated. Which of the following statements is true regarding these two types of actuators?
   A. These actuators are both rotary type and single-acting type.
   B. The one on the top is a scotch yoke actuator, and the one on the bottom is a rack-and-pinion type, but both actuators have linear movement.

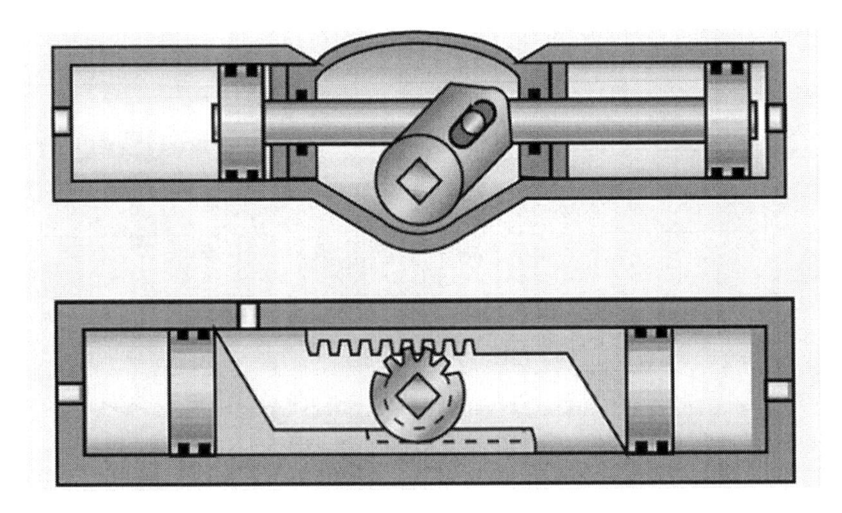

FIGURE 4.23 Two types of actuators.

C. Both actuators are double acting with linear movement.
D. The one on the top is a scotch yoke actuator, and the one on the bottom is a rack-and-pinion type; both have rotary type actuators, and they are both double acting.

Answer) On top, there is a scotch yoke actuator, while on the bottom, there is a rack-and-pinion actuator. As neither actuator has a spring, they should be operated by the power of the fluid supply, which can be either air or hydraulic oil, so they are double-acting actuators. Both types of actuators are rotary type, since the linear motion is converted into rotary motion by the yoke and connection pin on the scotch yoke actuator and by the pinion in the rack-and-pinion type actuator. Thus, option D is correct.

9. Which types of actuators could be the source of fugitive emissions?
   A. Hydraulic rack-and-pinion actuators
   B. Gas-over-oil as well as direct gas actuators
   C. Liner and rotary electrical actuators
   D. Pneumatic rack-and-pinion actuators

   Answer) The only source of fugitive emissions in actuators is gas. Hydraulic oil spills in the environment are harmful to the environment, but they are not regarded as sources of fugitive emissions. As far as the environment is concerned, air and electricity are 100% clean sources of energy. As a result, hydraulic rack-and-pinion actuators, linear and rotary electrical actuators, and pneumatic rack-and-pinion actuators do not emit fugitive emissions. The correct answer is option B since both gas-over-oil and direct gas actuators utilize gas, which is a source of fugitive emissions.

10. In relation to actuators, which of the following statements is incorrect?
    A. Compared to hydraulic actuators, pneumatic and electrical actuators are more environmentally friendly.
    B. In addition to rack and pinions, scotch yokes can also be pneumatic or hydraulic in nature.
    C. Failsafe closed actuators are closed by means of a spring force.
    D. The size of an actuator increases when a pneumatic actuator is converted to a hydraulic actuator for a specific valve.

    Answer) Option A is the correct choice because hydraulic actuators are less environmentally friendly than both pneumatic and electrical actuators. It is correct to select option B, since both rack and pinion as well as scotch yoke can be pneumatic or hydraulic. Option C is correct since a failsafe closed actuator is closed by a spring force. A hydraulic actuator is more compact than a pneumatic actuator for a particular valve, so option D is not correct.

11. In a failsafe closed hydraulic actuator, the oil pressure is very low, below the design minimum of 160 bar. In this situation, which of the following sentences is correct?
    A. The valve cannot be closed fully.
    B. The solenoid valve could be the reason for the low pressure of the oil.
    C. The valve is not getting opened fully.
    D. The exhaust valve could be less exposed to wear due to low oil pressure.

    Answer) A failsafe closed valve and actuator will close due to the spring force, so a low oil pressure should not cause any problems in closing the valve and actuator. The solenoid valve is not affected by the oil pressure. As a result, neither option A nor B is acceptable. Nevertheless, a low pressure of oil in the failsafe closed valve may prevent the valve and the actuator from fully opening. Therefore, option C is the correct answer. As the exhaust valve is not used in hydraulic actuators, option D is also incorrect.

12. In relation to pneumatic actuators, which of the following statements is incorrect?
    A. Different types of valves, such as isolation valve, check valve, solenoid valve, and exhaust valve, are located on a pneumatic actuator control panel.
    B. The components on the control panel could be connected with a tube or without a tube.
    C. Air filter and regulator can be integrated into one component on the control panel.
    D. The connection between the control panel and the actuator is tubeless.

    Answer) Except for option D, all of the options are correct. A small tube, such as 1/2", is typically used to connect the actuator to the control panel.

FURTHER READING

1. American Petroleum Institute (API). (2012). *Standard for actuator sizing and mounting kits for pipeline valves*. API 6DX. 1st ed. Washington, DC: API.
2. Arakelian, V., Le Baron, J. P., & Mkrtchyan, M. (2016). Design of scotch yoke mechanisms with improved driving dynamics. *Proceedings of the Institution of Mechanical Engineers, Part K: Journal of Multi-body Dynamics*, 230(4), 379–386.
3. Flowserve. (2018). *Valve actuation: The when, how, and why of actuator selection*. [online] available at: www.flowserve.com/sites/default/files/2018-05/(FLS-VA-EWP-00005-EN-EX-US-0518-Actuation_Type_Advantages_LR1.pdf [access date: 25th July, 2023]
4. Gonzalez, C. (2015). What is the difference between pneumatic, hydraulic and electrical actuators? *Machine Design*. [online] available at: www.machinedesign.com/linear-motion/what-s-difference-between-pneumatic-hydraulic-and-electrical-actuators [access date: 24th July, 2023]
5. Indelac Controls, Inc. (2014). *How to select an actuator: Comprehensive guide*. [online] available at: www.indelac.com/pdfs/How-to-Select-an-Actuator-Comprehensive-Guide.pdf [access date: 24th July, 2023]
6. International Organization of Standardization (ISO) 5210. (2017). *Industrial valves—multi turn valve actuator attachments*. 2nd ed. Geneva, Switzerland: ISO.
7. International Organization of Standardization (ISO) 5211. (2017). *Industrial valves—part turn actuator attachments*. 2nd ed. Geneva, Switzerland: ISO.
8. International Organization for Organization (ISO). (2011). *Petroleum and natural gas industries—mechanical integrity and sizing of actuators and mounting kits for pipeline valves*. ISO 12490. 1st ed. Geneva, Switzerland: ISO.
9. Mahl, T. (2009). The new generation of electrical actuators. *Valve World Magazine*, 14(4), 49–51.
10. Metalphote of Cincinnati. (2018). *How to select a valve actuator: Types of valve actuators, appropriate sizing, safety criteria, and more*. [online] available at: www.mpofcinci.com/blog/how-to-select-a-valve-actuator/#Function [access date: 25th July, 2023]
11. Mustalahti, P., & Mattila, J. (2018, September). Nonlinear model-based controller design for a hydraulic rack and pinion gear actuator. In *Fluid power systems technology*. Vol. 51968, p. V001T01A020. New York, NY: American Society of Mechanical Engineers.
12. Nesbitt, B. (2007). *Handbook of valves and actuators: Valves manual international*. 1st ed. Oxford, UK: Elsevier.
13. Onditi, J., & Carey, E. (2018). *Valve actuation: The when, how and why of actuator selection. A guide to actuators for upstream and midstream oil and gas applications*. Irving: Flowserve.
14. Pishock, D. (2016). *Choosing between a double acting and spring return actuator*. [online] available at: https://valveman.com/blog/double-acting-vs-spring-return-actuators/ [access date: 24th July, 2023]
15. Schlumberger. (2018). *LEDEEN hydraulic actuators and controls*. [online] available at: www.products.slb.com/valves/brands/ledeen/ledeen-hydraulic-actuators [access date: 24th July, 2023]
16. Sotoodeh, K. (2019). Actuator sizing and selection. *Springer Nature Applied Science*, 1, 1207, Springer. https://doi.org/10.1007/s42452-019-1248-z
17. Sotoodeh, K. (2021). *Prevention of valve fugitive emissions in the oil and gas industry*. 1st ed. Oxford, UK: Elsevier (Gulf Professional Publishing). ISBN: 978-0-323-91862-6
18. Sotoodeh, K. (2022). *Coating application for piping, valves and actuators in offshore oil and gas industry*. 1st ed. Boca Raton: CRC Press (Taylor and Francis). ISBN: 9781032187198

5 Safety Components

5.1 INTRODUCTION

The basic definition of safety is that it is the condition of being protected from physical, social, spiritual, financial, political, emotional, occupational, psychological, educational, and other types of failures, damages, errors, accidents, and harm or any other undesirable event. Safety can also be defined as the control of recognized hazards to achieve an acceptable level of risk. In most cases, it involves protection from the event or from exposure to something that may result in harm to one's health or financial loss. Aside from protecting people, safety components are also capable of protecting possessions. A variety of industries, such as oil and gas, require health, safety, and environmental engineering, also known as safety or loss prevention engineering, in order to minimize the risk and consequences of adverse events and hazards. Safety engineers are also responsible for ensuring that noise levels, emissions, water contamination, and other contaminants to the environment are within acceptable limits, taking legal and standard requirements into consideration. There are two important aspects of safety in oil and gas plants: process safety and personnel safety. It is possible for process safety hazards to cause serious accidents, such as the release of flammable, reactive, explosive, or toxic materials or the release of energy (such as fires and explosions). Safety engineering involves identifying hazardous failure modes that may lead to serious consequences, such as death, and setting targets for maximum tolerable frequencies for each mode of failure. Safety-critical equipment or components are specific pieces of equipment, control systems, or protective devices that, upon failure, could result in a hazardous situation that could result in an accident or could directly result in an accident affecting people and the environment. Safety-critical elements (SCEs) are parts of an installation, including its structure, equipment, and systems, whose failure could cause or contribute substantially to the release of a hazard with significant consequences. In the following section, various safety components and systems are discussed in more detail.

5.2 SAFETY COMPONENTS AND SYSTEMS

In this section, the most important safety components and systems are explained in alphabetical order.

5.2.1 Blowdown and Relief Systems

Oil and gas facility blowdowns are safety-critical operations, which must be performed in order to ensure the safe shutdown of processing facilities in the event of

DOI: 10.1201/9781003465881-5

an emergency. In the event of a full facility blowdown, the entire plant is isolated and depressurized into the flare system of the facility. Many benefits are associated with blowdown systems, including ease of maintenance and a reduction in maintenance time. In response to the activation of the blowdown valve, the blowdown system initiates a depressurization process. For maintenance or emergency purposes, blowdown valves depressurize a system or equipment by sending unwanted fluids to a flare. Basically, blowdown valves are emergency shutoff valves that operate in response to signals from emergency shutdown systems. The valve is normally closed, but in the event of an emergency, it will open in order to allow the fluid to be released.

When an overpressure occurs, it may be caused by a variety of factors, such as a process abnormality, fire, blocked discharge, excessive chemical reaction, or power outage. The term overpressure refers to an imbalance or disruption of the normal flow of material and energy. During abnormal conditions, a relief system is used to discharge gas from a pressurized vessel or piping system into the atmosphere in order to relieve pressures in excess of the maximum allowable working pressure (MAWP). In order to accomplish this, a manual or controlled pressure relief valve may be used, or an automatic pressure relief valve may be used.

5.2.2 Deluge Systems

Sprinklers are explained in more detail later in this chapter.

5.2.3 Emergency Communication Systems

During an emergency, communication can be challenging. There are many differences between communicating during an emergency and communicating during a normal period of time. It is possible for many aspects of a situation to change suddenly during an emergency. Consequently, an emergency communication system should be capable of performing a number of special functions. The term "emergency communication systems" refers to systems (typically computer based) that provide one-way and two-way communication of emergency information between individuals and groups. A number of different devices can be used to transmit information through these systems, including signal lights, text messages, live streaming video, televisions, and telephones. A closed-circuit television (CCTV) system is also considered part of an emergency communication system in some safety-related literature. An example of CCTV is a television system in which signals are not broadcast publicly but are monitored, primarily for the purpose of surveillance and security. A CCTV system relies on strategically placing cameras and viewing the camera's output on monitors. By design, closed-circuit cameras communicate with monitors and/or video recorders through private coaxial cable runs or wireless communication links, and, as a result, the content of the cameras is limited to those who are able to view them.

5.2.4 Emergency Escape Routes

Emergency escape routes should be designed in such a way that anyone confronted by fire anywhere within a building can escape to a place of reasonable safety. In this way, they will be able to reach a place of total safety outside the building immediately. During the time that the building is occupied, these routes and exits must be readily accessible. If there are any obstructions or alterations to these routes, it is important to be aware of them and report them if necessary. It is imperative that all changes in direction in corridors, stairways, and open spaces that are part of an escape route be clearly marked in order to ensure a safe exit from a building as soon as possible. Fire exit signs (also known as fire escape signs or emergency exit signs) indicate the direction of travel for safe departure from buildings. The next escape route sign should always be visible to an individual who is attempting to escape. Consistent spacing and placement of signs are essential. In addition to these routes and signs, all emergency lights are included in the fire protection system.

5.2.5 Emergency Shutdown Systems

An emergency shutdown system (ESD) is a highly reliable control system specifically designed for high-risk industries such as oil and gas, nuclear power, and other environments prone to explosions. An emergency shutdown system enables a process to be rapidly halted and isolated from incoming and outgoing connections in order to reduce the likelihood of an unwanted event occurring, continuing, or escalating. The systems are designed to protect personnel, equipment, and the environment if the process exceeds the control margins. Generally, ESD systems perform the following functions: shutting down part systems and equipment, isolating hydrocarbon inventories and electrical equipment, preventing hazardous events from escalating, halting hydrocarbon flow, depressurizing and blowing down, controlling emergency ventilation, and closing fire doors. ESD systems typically consist of a process transmitter, a logic solver, and a shutdown valve or blowdown valve. A transmitter monitors pressure, temperature, or level in process facilities such as pressure vessels and piping against a predefined limit and provides signals to a logic solver, which makes appropriate decisions based on the received signals. In response to an electrical signal generated by the logic solver, the final element takes an action, such as closing a valve to shut down a process system.

5.2.6 Fire Alarms

Our daily lives are enriched by the presence of fire alarms in offices, factories, and public buildings. A fire alarm system and fire alarms are often overlooked until an emergency occurs and they can save lives. No matter how the fire is detected, the occupants of the building will be alerted by a sounder that a fire is present and they must evacuate. Moreover, the fire alarm system may be equipped with a remote signal system, which would alert the fire brigade via a central station. Smoke, fire, carbon monoxide, or other emergencies can trigger a fire alarm system, which utilizes a number of devices to warn people through visual and audio signals. In the event of a malfunctioning piece of equipment, a process deviation, or abnormal conditions, an alarm system provides operators with audible and/or visual warnings. Through the monitoring of environmental changes associated with combustion, a fire alarm system detects the occurrence of fires. Various types of fire alarms are available, including those that are automatic, manually operated, or both. An oil and gas processing facility must be equipped with a fire alarm system that includes heat, flame, and gas detectors that are automatically activated; pull stations for manual activation; and strobe lights and sirens that will alert personnel to fires. Manually triggering an alarm can be accomplished by breaking a glass, pushing a button, or lifting and pulling a handle. In manual fire alarms, a variety of colors are used, but red is the most common color. National Fire Protection Association (NFPA) 72 covers the application, installation, location, performance, inspection, testing, and maintenance of fire alarm systems, supervisory station alarm systems, public emergency alarm reporting systems, fire warning equipment, and emergency communications systems (ECSs), as well as their components. Regular inspections of fire alarms are necessary to ensure compliance with the NFPA or other relevant fire standards. In the end, the effectiveness and reliability of a fire alarm system are determined by the choice of detection devices, their sensitivity, and the location of the detectors, as well as the materials and methods used in their installation and maintenance. In addition to providing occupants in the building with safety, fire alarm systems provide early notification of fires, reduce damage to buildings and plants, and reduce losses to humans and businesses. There are two types of alarms: visual (see Figure 5.1 on the right side) and audible (see Figure 5.1 on the left side). Alternatively, they may

FIGURE 5.1 Visual and audible (bell) fire alarms.

(Courtesy: Shutterstock)

be speaker strobes that sound an alarm, followed by an evacuation message that warns residents not to use elevators. During a fire, visual alarm devices (VADs) are used to alert deaf and hard of hearing individuals. To alert those who cannot hear the fire alarm sounder, pulses of flashing light are emitted. Visual alarms must be of a color, power, and operation that are clearly visible from all areas they cover, as well as easily visible from artificial light inside and sunlight outside. Audible alarm devices emit an audible warning signal when an alarm is triggered by an electrical or electro-mechanical device. A fire alarm should sound in a different tone from a combustible gas alarm and a toxic gas alarm. When an audible alarm is required, it is important to consider the number of different sounds required, as well as the number of sounds that plant personnel are capable of distinguishing. The use of voice messages allows for greater flexibility in conveying messages, but they may not be as effective or understandable to personnel working off-site or in multilingual environments. Silencing the alarm should require a manual operation. The alarm should not be automatically silenced or cancelled.

Figure 5.2 illustrates the fire alarm control panel or fire alarm control system as the brain of the system. All detectors are wired to the control panel, which also provides status indications to the user. There are various types of detectors connected to the fire alarm control panel, including smoke detectors, heat detectors, and flame detectors. A simulation of an alarm is also included in the control panel unit. Basically, fire alarm control panels receive signals from initiating devices such as detectors or manual call points and process these signals in order to determine the output functions. It is essential that the audible and visual alarms be compatible with the panels from which they are connected in order to have automatic activation from the fire and gas system panels. All employees should know what to do in the event of a fire alarm.

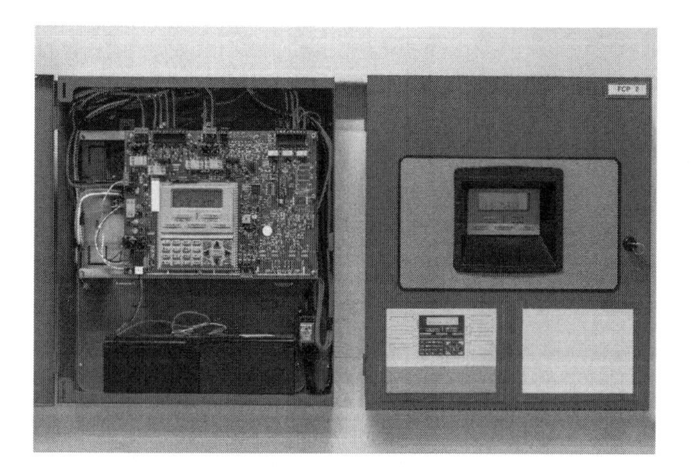

FIGURE 5.2 Fire alarm control system.

(Courtesy: Shutterstock)

There is a device known as a fire alarm call point or break glass call point (see Figure 5.3) that allows personnel to raise an alarm manually. Call points for fire alarms and break glass are fundamentally small boxes mounted on walls that contain buttons or levers that can be manually pushed to activate the alarm. In most cases, manual call points are located at exit fire doors. In order to ensure that manual call points are easily accessible, well illuminated, and conspicuous, they should be located 1.4 meters above finished floor level. It is recommended that they be placed against a contrasting background in order to make them more noticeable.

A fire alarm system may be conventional, addressable, intelligent, or wireless. A conventional fire alarm is an ideal solution for small buildings such as houses, offices, or retail shops. When a conventional fire alarm detects smoke in its immediate vicinity, it sounds, which is ideal for evacuating people from small areas. There is a physical connection between several call points and detectors in a conventional fire alarm system. In an addressable system, all detectors and manual call points are wired to the control panel. Detection is similar to that of a conventional system, except that each detector is assigned an address by means of a switch. A control panel can be used to determine which detector or call point triggered the alarm. There is a greater need for addressable fire alarms in large buildings and complexes of buildings. In contrast to conventional fire alarm systems, addressable fire alarms have a control panel that displays which devices are triggered by personnel or the fire department. The last type of fire alarm system (wireless) that will be discussed is the wireless fire alarm system. These effective alternatives can replace traditional wired fire alarm systems in all applications. The wireless communication between sensors and controllers is accomplished through the use of secure radio communications. The concept is a simple one that

FIGURE 5.3 Fire alarm call point.

(Courtesy: Shutterstock)

provides a number of unique advantages, including the fact that it is intelligent without the need for cabling.

5.2.7 Fire and Gas Detectors

5.2.7.1 Flame Detectors

A flame detector is generally intended to protect areas where fires are likely to develop rapidly. The flame detector detects the light emitted by flames or glowing embers in order to detect the radiation they produce. Flame detectors can be classified as three types: optical flame detectors that use infrared (IR), ultraviolet (UV), or a combination of both. There are a variety of hazardous environments where optical flame detectors can be used, including industrial heating and drying systems, industrial gas turbines, and petrochemical oil and gas facilities. By detecting fires as soon as possible, flame detectors can reduce the risk of an unwanted fire occurring in a given environment. Because of this, they are more reliable, especially in outdoor settings where they can detect flames more quickly than heat detectors or smoke detectors. Figure 5.4 illustrates a system for detection and monitoring of fire and gas in an oil and gas central processing platform through the use of infrared rays and alarms.

5.2.7.2 Gas Detectors

Gas detectors can be used to detect combustible, flammable, and toxic gases, as well as oxygen depletion. Devices of this type are widely used in industry to monitor manufacturing processes and firefighting systems, such as on oil rigs. Gas

FIGURE 5.4 A fire detection and monitoring system.

(Courtesy: Shutterstock)

detectors can either be fixed or portable. Fixed detectors are permanently installed at a specific location to provide continuous monitoring of plant and equipment. Portable detectors are small, handheld devices that are used to test an atmosphere in a confined space before entering, to track leaks, or to warn of the presence of flammable gases in a hazardous area. In situations where there is a possibility of a leak into an enclosed or partially enclosed space where flammable gases may accumulate, fixed detectors are particularly useful. Portable gas detectors can be divided into two types. The first type of portable gas detector is typically a small, handheld device capable of testing the atmosphere in confined spaces before entry, tracing leaks, or supplying early warnings when hot work is being performed in an environment that may contain flammable gases or vapors. Second, there is equipment that cannot be carried by hand but can be easily moved from one location to another. Gas detectors can be divided into two categories based on the way they measure flammable or toxic gases: point detectors and open-path detectors. For point detectors, the flammable gas detector measures the percentage by volume of flammable gases in the air using lower flammability limit (LFL) or upper flammable limit (UFL), while the toxic gas detector measures the concentration in parts per million (ppm). Open-path detectors, also called beam detectors, consist of a radiation source and a physically separate, remote detector. The detector measures the average concentration of gas along the path of the beam. Fixed gas detectors can be either point or open-path detectors, whereas portable gas detectors are always point detectors. In Figure 5.5, a portable and point gas detector is shown that is used to detect hydrocarbon gas leaks in order to prevent fires and explosions.

FIGURE 5.5 Using a portable gas detector in a plant.

(Courtesy: Shutterstock)

Generally speaking, there are three types of flammable gas detectors: catalytic, infrared, and open-path line of sight. When gas oxidizes during the catalyzed reaction between the gas and oxygen in air, it produces heat, and the catalytic gas detectors convert this temperature change into a signal proportional to the amount of gas present. There are two wire heating coils in catalytic gas detectors. The active coil is the detector, whereas the reference coil is the reference cell, which is usually identical and made of platinum. Gases that are combustible will be burned by the sensors when they are exposed to them. The electrical resistance between the reference and detector cells increases with increasing temperature. A number of advantages can be attributed to catalytic gas detectors, including their robustness, ease of operation and installation, long life expectancy at low replacement costs, proven technology, and high degree of reliability. The main disadvantage of this type of sensor is that it will become inactive when contaminated with chlorinated and sulfur compounds, as well as hydrogen sulfide. Furthermore, this device relies on oxygen to detect gases, and prolonged exposure to high concentrations of combustible gases may impair its performance.

Gas detectors that use infrared technology work on the basis of absorption of IR light. Using an infrared source, they illuminate a volume of gas within the measurement chamber. A portion of the infrared wavelengths are absorbed as the light passes through the gas, while others remain unattenuated. An optical detector and electronic systems are used to determine the amount of absorption as a function of gas concentration. The microprocessor calculates and reports the concentration of gas based on the absorption. Gas detectors of this type are characterized by their quick processing time, usually less than 10 seconds. Additionally, they are not susceptible to contamination, as opposed to catalytic gas detectors. A point infrared detector or an open-path infrared detector is used primarily for the detection of hydrocarbon vapors ranging from 0 to 100% by volume. Detectors do not require oxygen, cannot be poisoned, and are not ambiguous above the detection limit. However, they do not detect hydrogen and are inherently sensitive to pressure.

In order to detect gas over a long distance, a line-of-sight gas detector is commonly used. In open areas or along fence lines, the open path gas detector detects combustible levels of hydrocarbon gases. An open-path detector transmits a focused beam of infrared light from an infrared source to a detector located at a distance from the area being monitored. The source of infrared light in open-path sensors consists of a narrow beam that illuminates the area between the source and the detector. There is also the possibility of a mirror at the end of the path reflecting the beam back to the detector. Since certain gases absorb infrared light, it is possible to detect gas anywhere in the beam.

In oil rigs, chemical plants, and manufacturing facilities where toxic fluids and gases are stored and processed, gas detectors are commonly used. Some toxic gases emit a distinct odor, while others, such as carbon dioxide, have no odor and cannot be detected by humans. The smell of hydrogen sulfide, for instance, is similar to that of rotten eggs. In addition to being a colorless toxic gas, hydrogen sulfide can be fatal at levels as low as 100 ppm with prolonged exposure. High levels of toxic gases can pose serious health and safety risks, such as explosions, fires, and illness or death, in oil refineries, wastewater facilities, wells, tanks, and compressor stations. The concentration of a toxic gas in the air is measured using toxic gas monitors. The purpose of toxic gas detectors is to detect harmful gas levels in potential leakage areas in order to prevent leaks. As a general rule, gas detectors are programmed with pre-set levels that are based on the type and concentration of gas being monitored. An alarm is activated when the toxic gas levels exceed or fall below a specified level in the toxic gas detector, which contains a sensor that monitors the gas levels.

5.2.7.3 Heat Detectors

The purpose of heat detectors is to detect the emission of heat from a fire. Generally, a detector is activated by conventional currents of hot air or combustion products or by radiation. Like an electrical fuse, a heat detector operates in a similar manner. The detectors contain an electric alloy that is heat-sensitive. Upon reaching a certain temperature, the alloy turns from solid to liquid, triggering the alarm. When a fire occurs, the heat or thermal detector detects an increase in temperature. A heat detector may be classified as either a fixed-temperature detector or one that detects temperature changes over time. Both devices are activated by the heat of a fire incident. A fixed-temperature detector signals when a predetermined temperature is reached on the detection element. The rate of rise detectors will trigger a signal when the temperature rises above a predetermined level. As a result of fewer false alarms, heat detectors have a higher reliability factor than other types of fire detectors. Despite this, thermal detectors respond to fires more slowly than some other detection devices due to this means of activation taking some time to complete.

5.2.7.4 Smoke Detectors

Smoke detectors detect visible and invisible smoke generated by fires. In the event that a fire involves ordinary combustible materials, smoke detectors are used to provide early warning. Depending on the fuels involved and availability of oxygen, the type, volume, and density of smoke produced during the development of a fire will vary widely. As shown in Figure 5.6, a smoke detector alarms when a fire is detected.

Ionization and photoelectric smoke detectors are two types of smoke detectors that are commonly used. As a result of the positioning of radioactive material between two electrically charged plates, smoke alarms with ionization technology cause air to become ionized and current to flow between the plates, causing the air to become ionized. Smoke disrupts the flow of ions upon entering the chamber, which reduces the flow of current and activates the alarm. Photoelectric smoke detectors use light as a signal for fire detection. A light sensor is located in the chamber of the alarm. A beam of light is projected across this chamber by a light that shoots a straight line across the chamber. This causes smoke to deflect

FIGURE 5.6 Smoke detector.

(Courtesy: Shutterstock)

the straight line of light if it enters the chamber. Upon shifting its path, straight light enters a photosensor in another compartment of the chamber when deflected by smoke. This results in the alarm being activated when the light beam strikes the photosensor. In comparison to ionization smoke alarms, photoelectric smoke alarms are more effective and faster at detecting fires.

5.2.8 Fire and Gas Detection Systems

As fire and gas detection systems (FGDSs) include both fire and gas detectors as well as alarms, they are also referred to as fire and gas systems (F&Gs). There are a number of basic components of the system, such as field-mounted detection equipment, manual alarm stations, a logic unit for processing incoming signals, alarms, and human-machine interfaces (HMIs).

5.2.9 Fire Compartments

An area within a building or facility that is surrounded by fire barriers on all sides, including the ceiling and floor, constitutes a fire compartment, which could be a door, wall, or any other shape. Compartmentalization serves the purpose of limiting the spread of flames and smoke within a specified area of a building following the containment of a fire. This will allow occupants of a building a greater amount of time to evacuate safely and firefighting services a greater amount of time to extinguish a fire. Fire compartments may also be referred to as fire barriers, which are structures made of fire-resistant materials that serve as heat shields and prevent the spread of flames. A bund or dike around a storage tank serves as a fire compartment or a fire barrier to contain spills of flammable liquids.

5.2.10 Fire Dampers

A fire damper serves the primary purpose of preventing flames from crossing a fire barrier. It is common to install them near the floor or the wall, as appropriate, and they play an integral role in protecting the barrier they are intended to protect. Upon reaching a set temperature, fire dampers close and prevent flames from passing through. Fire dampers are installed in floors, walls, duct works, and partitions to ensure the integrity of the building during a fire.

5.2.11 Fire Doors

It is possible to use a fire door (sometimes referred to as a fire protection rating for closures) to reduce the spread of fire and smoke between different compartments of a structure, as well as to provide a safe exit from an establishment, as shown in Figure 5.7. This door has a fire resistance rating. If fire doors are left open in non-emergency situations, heat and smoke will enter escape routes and prevent people from escaping. The emergency exit sign shown on the fire wall close to the fire door.

5.2.12 Fire Extinguishers

Fire extinguishers are portable active fire protection devices filled with chemicals that are used to extinguish or control small fires, often in an emergency. An out-of-control fire, such as one that reaches the ceiling, endangers the user (i.e. no

FIGURE 5.7 Fire door.

(Courtesy: Shutterstock)

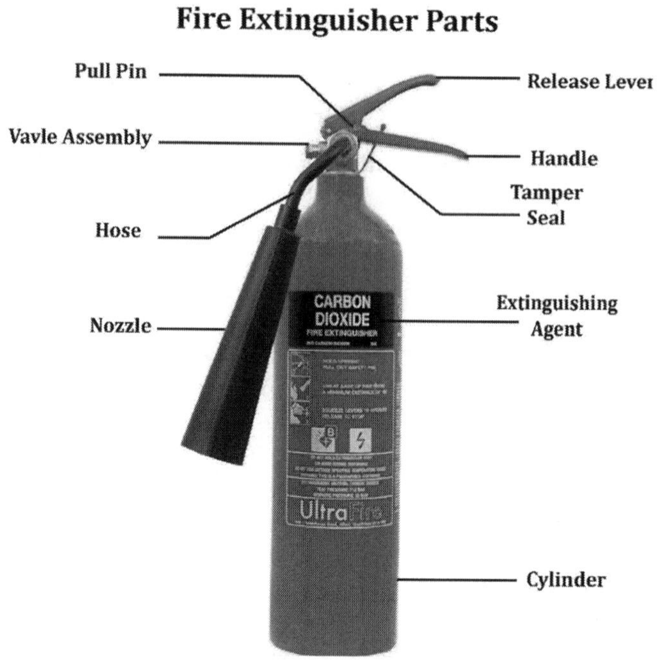

Fire Extinguisher Parts

Pull Pin
Release Lever
Vavle Assembly
Handle
Tamper Seal
Hose
Nozzle
Extinguishing Agent
Cylinder

FIGURE 5.8 A fire extinguisher and its components.

escape route, smoke, explosion hazards, etc.), or otherwise requires the assistance of a fire department, is not suitable for this device. A fire extinguisher, as illustrated in Figure 5.8, consists of the following main components:

Cylindrical tank: A cylindrical steel tank has a flat bottom, called the base, and a dome-shaped top, called the dome. Extinguishing agents and propellants are contained in these devices, which create the pressure necessary for the fire extinguishing agents to be sprayed.

Extinguishing agent: It is a substance that limits or suppresses the spread of fire. Knowing the types of fires (classes) and extinguishers that are needed is essential to selecting the appropriate extinguisher and extinguishing agent.

Valve assembly: Controlling or regulating the flow of the extinguishing agent is the function of the valve. A valve assembly consists of a body, a handle for lifting the fire extinguisher, a locking pin, a release lever, and a dip tube. There is also the possibility of finding a pressure gauge, or a pressure indicator, in the stored pressure fire extinguisher.

Nozzle/hose: Upon leaving the tank, a nozzle directs the extinguishing agent in the desired direction. Extinguishers weighing more than 3 kg are generally equipped with a hose, which allows the flow to be directed more precisely as it leaves the extinguisher.

Propellant: Extinguishing agents are expelled from fire extinguishers using a gas called a propellant. In stored pressure fire extinguishers, propellers are stored in the same chamber as the fire extinguishing agents. A cartridge-operated fire extinguisher contains the propellant in a separate cartridge, which is punctured when the extinguisher is activated, exposing the propellant to the extinguishing agent. Different propellants are used depending on the agent being used. In dry chemical extinguishers, nitrogen is typically used; in water and foam extinguishers, air is typically used.

The majority of fire extinguishers are hand-held cylindrical pressure vessels containing an extinguishing agent that can be discharged to extinguish a fire. Although they are less common, non-cylindrical pressure vessels can also be used to manufacture fire extinguishers. It is imperative that you understand the type of extinguisher you are using. There is a serious safety risk when the wrong type of extinguisher is used for the wrong type of fire. It is possible that spraying water on a grease fire can cause the grease to splatter, resulting in the fire spreading; similarly, dowsing live electrical equipment with water can result in an electric shock. It is the amount and type of agent contained in the extinguisher that determines the extinguisher's ability to extinguish a fire. There are certain advantages and disadvantages related to the use and limitations of extinguishers and extinguishing agents. As shown in Figure 5.9, there are five main types of fire extinguishers: water, carbon dioxide, dry chemical, foam, and wet chemical.

In the event of a fire involving solid combustibles, such as wood, paper and textiles, a stored pressure (fire class A) water extinguisher is the first type of fire extinguisher to be used. It is typically quite large and heavy and contains water under pressure, as is the case with the stored pressure water type extinguishers in red. In order to extinguish a fire, water fire extinguishers spray water from a spray nozzle, which helps to cover a larger area with water. An additional type of fire extinguisher is a carbon dioxide extinguisher, which can be used to put out fires of

FIGURE 5.9 Different types of fire extinguishers.

FIGURE 5.10 How to use a fire extinguisher.

classes B (flammable liquids) and C (flammable gases), as well as electrical fires. A dry chemical, dry powder, ABC powder, or multi-purpose fire extinguisher is a third type of extinguisher that is suitable for extinguishing fires of fire classes A, B, and C as well as electrical fires. Solid combustible and liquid flammable fires of classes A and B can be extinguished with foam fire extinguishers. To extinguish a cooking oil fire in a professional kitchen, wet chemical extinguishers are the only choice.

As shown in Figure 5.10, the following steps must be taken during using a fire extinguisher:

1. Pull the pin while holding the nozzle away from you to release the locking mechanism
2. Aim the hose nozzle low toward the fire
3. Squeeze the lever slowly to release the extinguishing agent
4. Sweep the nozzle from side to side at the base of the flames until they are extinguished

5.2.13 Fire Water Systems

The most important medium in the fight against fire is water. There is no doubt that water remains the most economical, most efficient, and most environmentally friendly method of extinguishing fires. The following is a list of the main equipment and facilities associated with fire water systems: The fire water storage tank is used to store the water. Depending on the location, underground or aboveground storage tanks can be used to store water. The pressurization and supply of fire water are accomplished by a variety of fire pumps. Pipes and fittings used in firefighting are primarily used for connecting firefighting equipment, conveying firefighting water, and so on. As piping components, pipe fittings facilitate the routing of pipes for directional changes, size changes, and branch connections. During inspection, testing, and maintenance, drainage systems must be installed in order to drain water from the equipment's intended use and also to allow for inspection, testing, and maintenance of the equipment. A fire in a storage tank is extremely difficult to extinguish. The only method that can effectively extinguish large fires in storage tanks is the use of foam systems. Typically, air foam is used for this purpose.

5.2.14 Hoses and Couplings

Fire hoses are high-pressure hoses that transport water or other fire retardants (such as foam) in order to extinguish fires. When it is used outdoors, it is usually attached to either a fire engine, a fire hydrant, or a portable fire pump. When used indoors, it can be permanently attached to the standpipe or plumbing system of the building. Generally speaking, the term "standpipe" refers to a rigid water pipe that can be built into buildings vertically or into bridges horizontally, to which fire hoses can be connected to apply water manually to fires. Within the context of buildings and bridges, standpipes serve the same purpose as fire hydrants. In general, fire hoses are made from non-perishable materials. A smooth bore with low friction is achieved by lining the bore with synthetic rubber, which is typically woven polyester. There is an option to apply an external coating to the hose in order to increase its resistance to abrasion. Polyvinyl chloride (PVC)/nitrile rubber is the coating on the outside of high-quality hoses, which forms a unified lining and outer cover. Polyvinyl chloride is a thermoplastic material that can be melted repeatedly at certain temperatures and then hardened. Through this process, PVC is made strong, lightweight, and durable. As shown in Figure 5.11, a fire hose is connected to a fire hydrant in order to extinguish a fire. In order to

FIGURE 5.11 Extinguishing a fire with a fire hose connected to a fire hydrant.

(Courtesy: Shutterstock)

connect firefighting hoses to hydrants, couplings are used. In order to extinguish large fires, fire hose nozzles must be used. By connecting them to a fire hose, they are able to distribute and direct pressurized water and fire retardants.

5.2.15 Hose Reels

There are several types of fire-fighting equipment, including hose reels, which are used as a first line of defense. Hose reels are cylindrical spindles made of metal, fiberglass, or plastic used to store hoses. It is important to note that hose reels are classified according to the diameter and length of the hoses they hold, the pressure rating, and the rewind method. A hose reel can either be permanently mounted, such as a wall-mounted hose reel, or it can be moved from one place to another. Fire hose reels are essentially hand-operated firefighting apparatus consisting of a reel with a water supply, a manual stop valve adjacent to the reel, and a semi-rigid hose connected to a shut-off device. Fires involving class A materials, such as paper, textiles, wood, most plastics, and rubber, are best handled with a hose reel. Fire hose reels provide a virtually unlimited supply of water in addition to being easy to operate and connected to the main water supply. They typically extend approximately 35–36 meters in length. Fire hose reels should not be used in electrical fires since water conducts electricity, leading to electroconduction. It is not uncommon for hose reels to be enclosed in a cabinet, as illustrated in Figure 5.12, in order to provide greater protection against damage and contamination. Some hose reels, however, do not have cabinets.

5.2.16 Hydrants

To put out a fire, a fire hydrant is essentially a pipe controlled by a valve that receives water from a water main. Hydrants are used to extract water from pipelines and distribution systems. It should be noted, however, that fire hydrants may

FIGURE 5.12 Fire hose reel inside a cabinet.

(Courtesy: Shutterstock)

also serve other purposes, including flushing fire water distribution systems, testing the water capacity of water distribution systems, and providing water. When firefighters are on duty, fire hydrants are visible points of connection to a water supply. In the event of a fire outbreak, fire hydrants provide quick access to water. In order to connect the trucks to the hydrants, hydraulic wrenches and standpipes are utilized. Each building, parking lot, roadside area, mine, industrial area, and so on must be equipped with a fire hydrant. Firefighters use these systems to provide them with the water necessary to extinguish and fight fires immediately. As early as the 18th century, fire hydrants were commonly used.

Figure 5.13 illustrates the main components of a fire hydrant. The vertical body of a fire hydrant is known as a barrel. It is this barrel that carries the fire water from the fire hydrant to the nozzles or outlets on the hydrant. There are three outlets on a traditional fire hydrant. In the middle of the cabinet, there is one main outlet in a larger size and two outlets on either side in a smaller size. The pumper truck is connected to the main outlet, while the hoses are connected to the side and smaller outlets. Pumper trucks are vehicles equipped with large tanks capable of pumping liquids and slurries into and out of them. It is common for firefighters to respond to fires with pumper trucks. Most fire hydrants are equipped with chains attached to their outlet caps that serve as a "leash" that ties the outlet caps to the hydrant in a series of three to six points. These caps must be

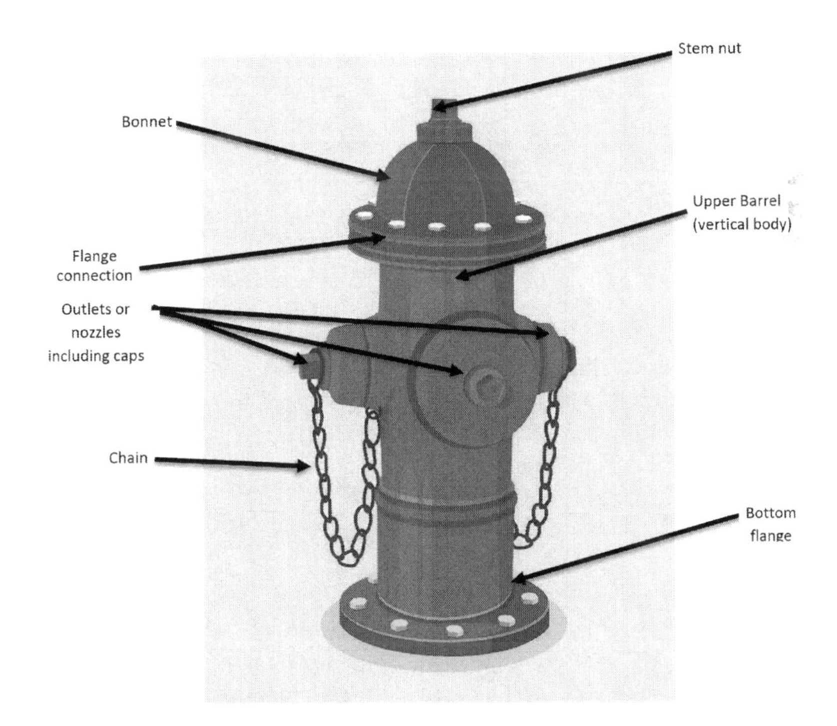

FIGURE 5.13 The fire hydrant and its components.

removed in order for firefighters to attach their hoses or suction tubes. You should avoid dropping the caps into the mud, snow, or weeds during the removal process, as their threads will become fouled and/or they will be difficult to locate once the hydrant has been placed back into standby mode. Besides preventing the caps from being lost or contaminated, the chains also prevent them from falling. Only the upper barrel is depicted in the figure, and there is an underground lower barrel that is connected to the upper barrel through its bottom flange. The bottom flange connection facilitates easy disassembly of the upper barrel from the lower barrel for inspection and maintenance purposes.

Fire hydrants have a conical cap called a bonnet that protects them from mechanical damage and water penetration. A flange connects the vertical body (barrel) and the bonnet. The term “flange joint” refers to a connection where the connecting pieces are connected by means of flanges (ring-shaped connectors) and bolted together. Hydrant bonnets are equipped with a stem nut for operating the valve inside and at the bottom. A shaft or stem is used to connect the stem nut on the top of the hydrant to the valve. When turned with a hydrant wrench, the stem nut turns the operating stem of the hydrant and raises the valve to the open position. It is necessary to open and close the valve located at the lower section of the hydrant in order to initiate or stop the flow of water inside the body or barrel of the hydrant. There is no provision in fire hydrants for throttling the flow of water; instead, the valves are designed to be operated either fully on or fully off. Fire hydrants typically have two types of valves installed on the lower portion: wedge gate valves and butterfly valves. As soon as the barrel hydrant is opened, the valve at the bottom rises, plugging the drain holes and simultaneously filling the barrel with water. The valve lowers once it is in its closed position, blocking water passage and reopening the drain holes at the bottom. An underground fire hydrant’s lower section contains the lower barrel, the main valve, and the drain plug or drain hole, as well as the base. Through the drainage holes, water is allowed to slowly drain from the hydrant barrel, preventing freezing.

Wet barrels and dry barrels are the two main types of fire hydrants. It is important to note that dry-barrel hydrants do not contain any water until the valve is opened. The purpose of this type of hydrant is to prevent problems caused by freezing water inside the barrel. In contrast to dry-barrel hydrants, wet-barrel hydrants enclose water when not in use. In areas where freezing water is not an issue, this type of hydrant is perfectly acceptable. A hydrant’s capacity can sometimes be determined by the color of its bonnet or by the markings on its exterior.

5.2.17 Lightning Protection

In lightning discharges, electrical charges build up in the atmosphere and are discharged in a massive manner to the earth. Lightning is without a doubt one of nature’s most beautiful creations. Lightning strikes, however, can cause devastating accidents in the oil and gas industry when they strike flammable materials. During lightning, people may be exposed to explosions caused by flammable materials. The purpose of lightning rods or lightning conductors is to protect structures from lightning strikes. If lightning strikes the structure, it will be conducted to ground through a wire rather than passing through the structure, which may cause a fire or electrocution. Through the use of a low-resistance wire, electrical energy is transferred directly to the earth by earthing or grounding.

5.2.18 Rupture Disks

Rupture disks are non-reclosing pressure relief devices used to protect pressure vessels, equipment, or process piping systems from overpressurization or dangerous vacuum conditions. There are two types of rupture discs: those that are used as single pressure protection devices and those that are used in conjunction with pressure safety valves. If the pressure safety valve fails to operate or is incapable of relieving excess pressure quickly enough, the rupture disc will spring into action and burst, controlling the pressure. Figure 5.14 shows an image of a ruptured disk before and after the overpressure has been relieved.

5.2.19 Safety Instrumented Systems

Sensors, logic solvers, and final control elements make up safety instrumented systems (SISs), which are all designed to bring the process to a safe state in the event that predetermined conditions are violated. In safety instrumented systems, plant parameters and values are monitored within their operating limits, including pressure, temperature, and fluid level. Alarms are activated when risk conditions arise, placing the plant in a protected state or even shutting down.

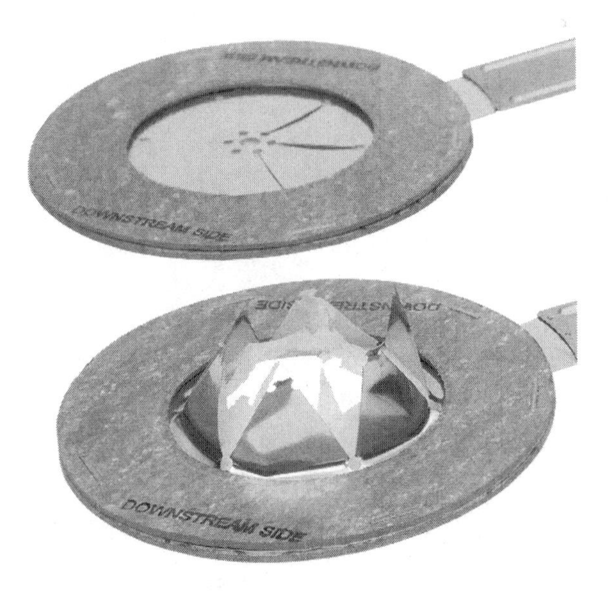

FIGURE 5.14 An image of a ruptured disk before and after relief of the overpressure.

5.2.20 Safety Valves

The purpose of safety valves is to ensure the safety of the user. The purpose of these valves is not to control the pressure in a system but to allow it to be released immediately in the event of an emergency or system failure. The purpose of both pressure safety valves and relief valves is to relieve pressure from a pressurized system, but their technical definitions are somewhat different. Relief valves are not used to control the pressure in a system, most commonly in fluid or compressed air systems. Instead of controlling the pressure in a system, these valves are designed to release pressure immediately in the event of an emergency or system failure. They open in proportion to an increase in system pressure. Safety valves, in contrast to relief valves, are designed to open quickly and completely in the event of a disaster rather than to release the pressure gradually. The most common type of device used to prevent overpressure in plants is a safety valve. Upon reaching a predetermined maximum pressure, the safety valve releases a volume of fluid from within the plant, thereby reducing the excess pressure safely. Most safety valves are spring loaded. A spring force closes the valve when the pressure drops to a certain level. Given that the safety valve may be the only device available to prevent catastrophic failure under overpressure conditions, it is imperative that it remain operational at all times. Figure 5.15 illustrates how a PSV is installed in a power plant's piping system.

5.2.21 Sprinklers

In firefighting engineering, a sprinkler system is a system of pipes and sprinklers used to control or extinguish fires. It includes a reliable and adequate water supply

FIGURE 5.15 A PSV in a power plant.

(Courtesy: Shutterstock)

as well as an interconnected network of pipes and sprinklers of specialized sizes. The system also includes a device that activates an alarm when it is operating. It is the responsibility of sprinklers to spray water forcefully over the flames in order to extinguish them completely or, at the very least, control the heat and limit the development of toxic smoke until the fire department arrives. Fire sprinklers are activated only when they are closest to the fire. The sprinklers are not always activated by smoke as they are by heat. The installation of fire sprinklers is a simple, effective, and economical process. More than 40 million sprinkler heads are installed each year throughout the world as a result of their widespread use. Sprinklers alone were responsible for controlling more than 96% of fires in buildings protected by fire sprinklers. Some countries require sprinklers to be installed in public areas such as hotels, factories, schools, meeting rooms, and museums. The function of sprinklers can be better understood if we take a closer look at them. An essential component of any water supply system is the sprinkler head. Figure 5.16 illustrates the components of the sprinkler, which is also called the sprinkler head. It consists of a plug, a liquid-filled glass bulb or fusible link, a sprinkler, and a frame. The fire causes the temperature in the environment to rise, causing the liquid in the glass bulb to expand. As the temperature rises, the liquid expands to the point where it breaks the glass, causing the water to flow out. Upon reaching the sprinkler, the water spreads over a larger area. Sprinkler systems can either be wet or dry, similar to fire hydrants. A wet pipe sprinkler system, also known as a traditional fire sprinkler system, is the most common type of sprinkler system. In this system, the sprinkler piping is constantly filled with water. A deluge sprinkler system is one type of sprinkler system where all sprinklers open simultaneously and the piping is unpressurized and dry. As a result of the use of open sprinklers, water will be distributed throughout the entire

FIGURE 5.16 A sprinkler and its parts.

(Courtesy: Shutterstock)

area. As the sprinkler pipe is filled with atmospheric pressure, a deluge valve is activated in the event of a fire by receiving a signal from a fire alarm or detector, allowing water to flow through the pipe from the water supply piping system. A deluge system is used in areas where there is a risk of rapid fire spread. Deluge systems provide a faster fire suppression response by releasing water simultaneously through all of their open heads in the event of a fire. It is necessary to turn on each sprinkler head individually when a sprinkler system has closed heads. Due to this, the deluge system is more effective in high-hazard environments.

QUESTIONS AND ANSWERS

1. Choose the correct sentence regarding fire hydrants.
   A. There are several components that make up a fire hydrant, including the hose, the operating nut, and the safety chain.
   B. A dry hydrant always contains water in the upper barrel.
   C. The valve below the frost line is activated by a nut located on the top cap of the hydrant.
   D. Fire hydrants have a vertical body known as a body. Fire water is transported from the fire hydrant to the nozzles or outlets on the fire hydrant via this body.
   E. Hydrants are protected from mechanical damage and water penetration by a conical cap known as a bonnet.
   F. A stem nut is provided on the flange of the hydrant in order to operate the valve from both the inside and the bottom.
   G. Fire hydrants cannot provide quick access to water during a fire outbreak.

   Answer) It is correct to select options A, C, and E. Option B is incorrect due to the fact that dry hydrants do not always have water in the upper barrel. Choosing option D is incorrect since the vertical body of the hydrant is called a barrel, not a body. The option F is incorrect since the fire nut is located on the cap rather than on the flange. Additionally, it is incorrect to choose option G since fire hydrants can provide quick access to water in the event of a fire.

2. Fill in the gaps with the correct words about safety valves and overpressure protection systems.
   Safety valves are designed to ensure the ______ of users. Both pressure safety valves and relief valves serve the purpose of _______ from a pressurized system, but their technical definitions are somewhat different. In the event of an emergency or system failure, ________ are designed to release pressure immediately instead of controlling the pressure within a system. When the system pressure increases, they open ________. As opposed to relief valves, safety valves are designed to open ________ in the event of a disaster rather than gradually releasing pressure. As

soon as a predetermined maximum pressure is reached, the safety valve releases a volume of fluid within the plant, thereby reducing the excess pressure in a safe manner. The majority of safety valves are _______.

A. Safety, relieving pressure, relief valves, proportionally, quickly and completely, spring loaded
B. Safety, relieving pressure, relief valves, proportionally, proportionally, actuated
C. Health, relieving temperature, safety valves, suddenly, quickly and completely, spring loaded
D. Health, relieving temperature, safety valves, suddenly, proportionally, actuated

Answer) Option A is the correct answer.

3. Find the correct statement about sprinkler systems.
   A. A sprinkler system is a system of pipes and sprinklers used to control or extinguish fires.
   B. A sprinkler system requires a reliable and adequate supply of foam as well as an interconnected network of pipes and sprinklers of specialized sizes.
   C. A sprinkler system cannot be connected to a fire alarm system.
   D. The installation of fire sprinklers is a simple, effective, and economical process.
   E. In the event of a fire, deluge systems provide faster fire suppression than sprinkler systems by simultaneously dispersing water through all of their open heads.

   Answer) It is correct to select options A, D, and E. It is incorrect to select option B since sprinklers supply water rather than foam. Option C is also incorrect since sprinkler systems can be connected to fire alarms.

4. What is the name of the safety system which consists of a process transmitter, a logic solver, and a final element?
   A. Emergency shutdown system
   B. Fire and gas detection system
   C. Corrosion protection system
   D. Emergency escape route

   Answer) Option A is the correct answer.

5. What is a one-time use safety relief device?
   A. Safety valve
   B. Rupture disk
   C. Relief plug
   D. All options are correct

   Answer) Option B is the correct answer.

6. Which type of safety-critical system uses rods and conductors?
   A. Emergency communication
   B. Corrosion protection
   C. Lightning protection
   D. Firefighting protection

   Answer) Option C is the correct answer.

7. Firefighters use which fire safety component to provide them with the water necessary to extinguish and fight fires immediately?
   A. Sprinkler
   B. Hydrant
   C. Deluge
   D. Fire door

   Answer) Option B is the correct answer.

8. What safety system does CCTV belong to?
   A. Emergency shutdown system
   B. Safety instrumented system
   C. Emergency communication system
   D. Fire and gas detection system

   Answer) Option C is the correct answer.

9. Which safety component is located close to the emergency exit?
   A. Fire door
   B. Fire extinguisher
   C. Fire damper
   D. Safety valve

   Answer) Option A is the correct answer.

10. What components can be used in conjunction with hoses?
    A. Couplings
    B. Hose reels
    C. Nozzles
    D. All three choices are correct

    Answer) Option D is the correct answer.

FURTHER READING

1. BRE Trust. (2014). *Visual alarm devices—their effectiveness in warning of fire.* [online] available at: www.bre.co.uk/filelibrary/Briefing%20papers/VAD-Briefing-Paper.pdf [access date: 29th July, 2023]
2. Bryan, A., Smith, E., & Mitchel, K. (2013). *Fire and gas systems engineering handbook.* Columbus, OH: Kenexis Consulting Cooperation.
3. Desai, D. J. (2019). *Study on firefighting system.* Texas: Vapi Emergency Control Center.

4. Galvan, A. (2007). A technical basis for guidance of lightning protection for offshore oil installation. *Journal of Lightning Research*, 3, 1–9.
5. Hind, J. (2009). *Fire and gas detection and control in the process industry*. Risk and Safety Group, Worley Parsons. [online] available at: https://www.scribd.com/document/165003757/Fire-and-Gas-in-the-Process-Industry-Jon-Hind-Paper
6. Iranian Petroleum Standard (IPS) E-SF-260. (2010). *Engineering standard for automatic detectors and fire alarm systems*. Tehran, Iran: IPS.
7. Kletz, T. A. (1990). *Critical aspects of safety and loss prevention*. 1st ed. Oxford, UK: Butterworth-Heinemann. ISBN: 978-0-408-04429-5
8. National Association of Fire Protection (NFPA) 22. (2022). *Standard for water tanks for private fire protection*. Cambridge, MA: NFPA.
9. National Association of Fire Protection (NFPA) 72. (2013). *National fire alarm and signalling code*. Cambridge, MA: NFPA.
10. National Association of Fire Protection (NFPA) 291. (2013). *Recommended practice for water flow testing and marking of hydrants*. Cambridge, MA: NFPA.
11. Nolan, P. (2011). *Handbook of fire and explosion protection engineering principles*. 2nd ed. Oxford, UK: Elsevier. ISBN: 978-1-4377-7857-1
12. Nolan, D. P. (2019). *Handbook of fire and explosion protection engineering principles for oil, gas, chemical, and related facilities*. 4th ed. Oxford, UK: Elsevier. ISBN: 978-0-12-816002-2
13. Shrivastava, N., & Shukla, V. (2019). Fire detection & alarm system in oil & gas refinery. *International Journal of Scientific Research & Engineering Trends*, 5(1).
14. Wise Global Training Ltd. (2015). *Introduction to oil and gas operational safety revision guide for the NEBOSH international technical certificate in oil and gas operational safety*. 1st ed. New York, NY: Routledge. ISBN: 9780415730778

6 Flare

6.1 INTRODUCTION

In order for a process plant to operate and function safely and satisfactorily, the flare system is one of the most important elements for operational and emergency relief of flammable substances in the liquid or gaseous phases. Most people view flares as merely a fire on top of a pipe that burns gases. Flares are often perceived by the public as a source of odor, smoke, noise, fallout, and light. Gas flaring is the process of burning natural gas as part of the oil extraction process. It has been practiced since the beginning of oil production over many years ago. In many industries, significant amounts of waste streams, such as hydrocarbon vapors, must be disposed of continuously or intermittently. Examples of off-spec products or bypass streams generated during start-up operations can be provided. Due to safety and environmental concerns, direct discharge of waste gas streams and vapors into the atmosphere is unacceptable. As a standard practice, gas flaring is designed to convert flammable, toxic, and corrosive vapors into environmentally acceptable waste products. By means of combustion, gas flaring converts flammable, toxic, or corrosive vapors into less objectionable compounds. Many plants require that flares be designed according to strict safety principles as they are a critical operation. Flaring systems typically consist of equipment that safely burns vented hydrocarbons at a pressure drop that does not compromise plant relief mechanisms. An ideal operating condition would be to eliminate the need for flares, since they waste hydrocarbons that could be converted into products and thereby increase profits. However, facilities recovering large amounts of released hydrocarbons under emergency conditions is currently not economically justified.

It is important to note that the flare function is primarily a safety function. Flare systems protect equipment against pressure build-ups that could result in explosions. Additionally, the flare system collects the "fatal gases" and burns and vents them into the atmosphere. It is generally recommended that special consideration be given to the design of various safety facilities in order to prevent catastrophic equipment failure when planning and laying out process plants. In addition to preventing overpressure, these facilities facilitate the safe disposal of discharged vapors and liquids. As part of the operation of these facilities, portions are also used for safe disposal of hydrocarbons—particularly during the start-up and shutdown phases. As a result of local regulations or plant practices, direct discharge of waste or excess vapors into the atmosphere is not acceptable. There is a possibility that the concentration of contaminants at ground level or on adjacent platforms may exceed the permissible explosion or toxicological threshold limits. Conditions that may be hazardous may arise due to weather considerations, such

DOI: 10.1201/9781003465881-6

FIGURE 6.1 Gas flaring in an offshore oil and gas platform.

(Courtesy: Shutterstock)

as severe temperature inversions of extended duration. A range of factors contribute to flare-ups, including market and economic constraints as well as a lack of appropriate regulation and political will. As for the economic and technical reasons for gas flaring, it is important to note that oil fields are often located in remote and difficult-to-access locations. These sites may not produce consistent or large amounts of associated gas that can be utilized by operators. It can be logistically and economically challenging to transport associated gas to a location where it can be processed and utilized due to this factor. Furthermore, capturing and utilizing associated gas is often perceived as prohibitively expensive if oil production sites are small and dispersed over a large area. Typically, associated gas is flared in these situations. The capture and utilization of associated gas may be economically and technically feasible; however, a country's laws and regulations may make it difficult or even impossible for companies to commercialize associated gas. As an example, an oil company may have secured the rights to extract oil but may not be able to utilize the associated gas produced during extraction. The regulations may not specify how associated gas should be handled commercially in other cases. Due to this, there is legal ambiguity regarding how associated gas should be treated. Figure 6.1 shows gas flaring in an offshore oil and gas platform. In general, non-hazardous vapors, such as low-pressure steam, can be discharged directly into the atmosphere. However, hydrocarbon vapors that are discharged continuously or intermittently cannot be discharged directly to the atmosphere. They should be disposed of in a closed system and burnt in a flare.

6.2 GAS FLARING ENVIRONMENTAL IMPACTS

The natural gas that is not combusted by a flare is vented into the atmosphere as methane Approximately 139 billion cubic meters of gas will be burned by thousands of gas flares at oil production sites worldwide in 2022. Using a typical associated gas composition, flare combustion efficiency of 98%, and global warming potential of 28 for methane, each cubic meter of associated gas flared results in approximately 2.6 kilograms of CO_2 equivalent emissions (CO_2), resulting in approximately 350 million tons of CO_2 emissions each year, of which approximately 42 million tons are unburnt methane emissions. As a result of the inefficient combustion of flares, methane emissions contribute significantly to global warming. As a warming gas, methane is over 80 times more powerful than carbon dioxide on a 20-year timeframe, according to the Intergovernmental Panel on Climate Change. In this case, the annual CO_2 emissions from flaring amount to approximately 126 million tons. Obviously, flares are unproductive and can be avoided much more easily than many other sources of greenhouse gases (GHGs).

6.3 FLARE TYPES

In general, there are two types of flare systems: *elevated flares* and *ground flares.* An elevated flare system uses a burner and igniter located at the top of a pipe or stack to carry out combustion reactions. During the process of releasing gases, they are sent through an elevated stack from a closed collection system in order to be burned off at the top. A ground flare is similarly equipped, except that the combustion takes place at or near the ground surface. A number of oil and gas sites are also subject to height restrictions if they are located in the vicinity of aircraft flight paths. As a result, operators are unable to safely flare excess natural gas without the use of ground flares. In other cases, a ground flare is the preferred choice only if a plant is located in an area where it is highly desirable to have a flare that is not visible to the public. The flame generated in a ground flare could be open or closed. Enclosed ground flares are when the flare is surrounded by an enclosure that shields it from noise and radiation. For industries located in urban areas, enclosed ground flares are the most practical type of ground flare. Through an opening at the top of the refractory-lined enclosure, hot combustion gases are discharged into the atmosphere. The term "open ground flare" is used when large piping systems are used to distribute released gases and liquids to flares on the ground, and the area is surrounded by a radiation fence. While they are expensive, they provide smokeless combustion without a visible flame.

There are several factors that will influence the selection of a flare, including the availability of space; the characteristics of the flare gas (i.e. composition, quantity, and pressure level); economics, including the initial investment and operating costs; and concerns over public relations with the community surrounding the facility. It is advisable to place flares at a considerable height above the ground in order to ensure safety. Due to their lower costs, elevated flares are preferred over enclosed ground flares. Furthermore, elevated flares are preferred over open ground flares due to their lower land requirements. The following are the principal advantages of a ground flare system: There is no need for structural support. The erection process is relatively straightforward and requires only a few light parts. It is easy to maintain. Because the flare is enclosed in a box, the flame cannot be seen (this is only true about closed ground flares). Finally, in general, it is a fairly quiet system. As a disadvantage, ground flares must be well isolated from the remainder of the plant and process lines, thereby requiring a considerable amount of space and long interconnecting piping. As a result of combustion occurring at ground level, toxic gases may be present in relatively high concentrations. An elevated flare, on the other hand, requires less ground space. As a result of its high elevation, it can be located within a process area or on the perimeter of the plant site. This can result in lower piping costs due to shorter and smaller pipe runs. Additionally, the distance between the point of discharge from safety valves and the flare stack is shorter than for ground flares. It is also a problem with elevated flares that maintenance is tedious and time consuming. Visibility of the flame is the most serious disadvantage, and it sometimes causes objections from local residents. Additionally, these systems require a greater amount of steam in order to produce a smokeless flare. There is also the disadvantage of relatively high noise levels. It is generally recommended to use elevated flares. While ground flares have made numerous advances, their need for a large area of land and their high initial cost make them less attractive than elevated systems. It should be noted, however, that visibility of the flame, depending on local regulations, may also be a determining factor in some cases.

6.4 FLARE COMPONENTS

The flare system, as illustrated in Figure 6.2, consists of the following components:

- Set of depressurization components (safety valves, rupture disks, blowdown valves [BDVs], thermal safety valve, automatic pressure control valves)
- The main collection network and one or two secondary collections (also known as manifolds)
- A separator drum at the foot of the flare that separates the various phases of hydrocarbons (water, liquid, and gaseous)
- A sealing device that prevents air from entering the system (purge gas, hydraulic guard)
- Flare stacks are placed on top of flare tips when a flare is lit
- A pilot light gas supply system is installed to permanently supply the pilot lights located close to the flare tip (flare ignition system)
- An ignition system is installed for these flares
- Controls and instrumentation

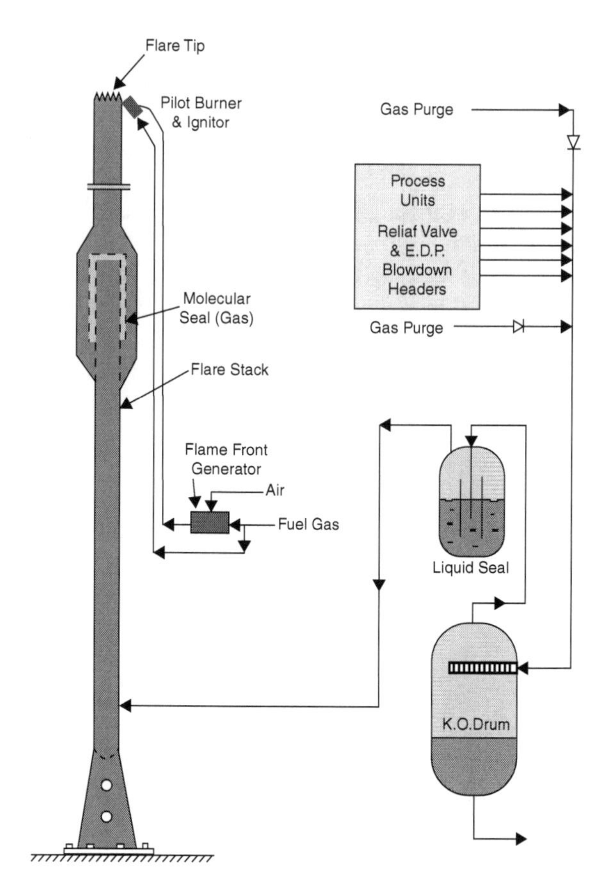

FIGURE 6.2 A flare system and components.

6.4.1 Safety Valves

Safety is a concern in any system containing potential energy, whether electrical, chemical, or fluid. Each has methods of protection, and for fluid systems, certain shutoff valves are used in situations when excess pressure can be a problem. The function of a safety valve is to act as a failsafe. One example of a safety valve is a pressure relief valve, which automatically releases a substance from a boiler, pressure vessel, or other system when pressure or temperature exceeds preset limits. The purpose of a safety valve is to prevent overpressure in a system. As a result of overpressure, the system's pressure exceeds the maximum allowable working pressure (MWAP) or the pressure for which the system is designed. In comparison with relief valves, safety valves can open very quickly. A safety valve works by opening gradually, then fully, allowing the unwanted pressure to be removed from the system as quickly as possible. Safety valves prevent pressure increases that can cause malfunctions, fire hazards, or explosions. It is the system's media that fully

FIGURE 6.3 A pressure safety valve.

(Courtesy: Shutterstock)

activate a safety valve in the event of a power outage, ensuring its continued operation. When electronic or pneumatic safety devices fail, safety valves are the only devices that operate mechanically. In Chapter 3, there is more information about safety valves. The operation of each pressure relief valve should be examined individually for any probable causes of overpressure, including operational failures and plant fires. It is imperative that the valve be sized according to the case that will require the maximum amount of relief. It is possible to provide two separate safety valves, one for the fire condition and one for operational failure, if there is a fire condition that is controlling. A pressure safety valve is shown in Figure 6.3.

6.4.2 Rupture Disks

An rupture disc is composed of one-time-use membranes that fail at a predetermined differential pressure, either positive or negative. When over- or under-pressurization occurs in process piping systems, rupture discs respond immediately. However, once the disc ruptures, it is not resealed. A rupture disc has several major advantages over pressure relief valves, including failsafe performance, leak proofing, and being the most economical (cost-wise). The rupture disc is usually round in shape, has one or more layers, and can either be flat or curved (see Figure 6.4). For different applications, different rupture discs are required. In addition to stainless steels, higher-quality materials such as Inconel, Hastelloy, and Tantalum are used for rupture discs. Discs can be installed directly between flanges or in a rupture disc holder, which is mounted between the flanges.

As the name implies, rupture discs are non-reclosing pressure relief devices used to prevent overpressurization of pressure vessels, equipment, or process

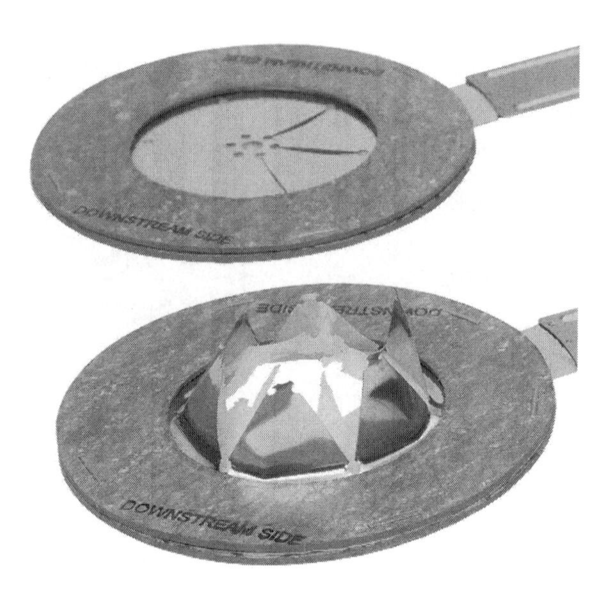

FIGURE 6.4 An image of a ruptured disc before and after relief of overpressure.

piping systems. It is possible to use the rupture disc either independently or in conjunction with a pressure safety valve in order to provide pressure protection. Occasionally, in cases of overpressurization, if the pressure safety valve does not operate or cannot relieve excess pressure fast enough, a rupture disc will burst, releasing the pressure. When a rupture disc is installed before or upstream of safety valves, it can assist in protecting the functionality and reliability of the valves when the process piping system contains corrosive, adhesive, polymerizing, or viscous fluids. Consequently, less expensive pressure safety valve materials can be used with corrosive fluids. It is also less expensive to replace a rupture disc than to replace a pressure safety valve. In addition, safety valves can be tested without having to remove them from the process piping. By pressurizing the space between the rupture disc and the valve stroke, this can be achieved. As a result of the rupture disc's almost double back pressure resistance, it remains undamaged during the inspection of the safety valve.

6.4.3 Blowdown Valves

When blowdown valves are in the open position, they are used for operation. The purpose of blowdown valves is to depressurize a system or piece of equipment for maintenance or emergency situations by sending unwanted fluids to a flare. Blowdown valves are primarily used to initiate a blowdown in piping and pipeline systems when they are activated. In essence, blowdown valves are emergency-activated on-off valves that function in conjunction with emergency shutdown systems. Blowdown valves are primarily used to initiate a blowdown when they are activated in piping or pipeline systems. A downstream isolation valve is required for BDVs, as well as a bleed between the two valves. As part of remedial work, the downstream isolation valve will be closed once the pressure has been relieved, and a blind will be installed upstream of the BDV. During maintenance of the BDV, the upstream equipment or system will be depressurized, the downstream isolation valve will be closed, the BDV will be removed, and blind flanges will be installed on each open-ended flange. Until the BDV has been reinstalled and certified as functional, the system will remain shut down.

6.4.4 Thermal Safety Valves

A thermal relief valve may also be referred to as a thermal safety valve, a temperature relief valve, or a thermal expansion relief valve. It is a safety device employed in liquid piping and pipeline systems to protect equipment and systems. When the thermal expansion of a liquid creates excessive pressure inside a closed system, the thermal relief valve pops up to release some fluid and bring down the pressure back to an acceptable limit. The thermal safety valve, as the name implies, is for cases when thermal expansion of a liquid would create excessive pressure in a closed system. When liquid heats up, it expands, but not nearly as much as a gas. Even if the temperature increase is quite dramatic, the change in the volume of the liquid is not by a large percentage, nor is it usually very rapid. The more common industry term for this type of valve is "thermal relief valve." These factors lead to thermal valves with unique properties. First, they do not need to be very large, since the volume increase will be relatively small. Usually they have an inlet diameter no larger than 1". Liquid is also unsafe to vent directly into the environment, obviously in the case of petroleum products or acids. Even water that has become extremely hot can be an environmental hazard, so some sort of containment method is also critical near these thermal safety valves. TRVs work in a similar manner to pressure relief valves. In spite of the name thermal safety valve, this valve is primarily driven by pressure increases. Under normal operation, the TRV remains closed by spring force. When the fluid expansion force is significant enough and exceeds the internal spring force, the valve will open. Once the pressure reduces, the spring returns to close the thermal relief valve to work smoothly.

6.4.5 Pressure Control Valves

The purpose of a (pressure) control valve is to regulate or manipulate the flow of fluids such as gas, oil, water, and steam. An example of a final control element, it is a critical part of a control loop. In the modern industrial environment, the control valve is by far the most common final control element. An actuated control valve is illustrated in Figure 6.5. Generally, these are process control valves (globe valves with actuators) that are activated by an electronic, pneumatic, or hydraulic system so that excess fluid can be discharged to a flare either permanently or intermittently. Process control valves discharge to the blowdown circuit, but they are not included in this system.

FIGURE 6.5 A control valve.

(Courtesy: Shutterstock)

A valve actuator is the device connected to the valve stem that provides the force required to move the valve. There are several ways in which the actuator can be operated, including electrically, pneumatically, and hydraulically. As far as control valves are concerned, pneumatic actuators and diaphragm-type actuators are the most commonly used and most reliable. In Chapter 4, you will find more information about actuators. Control valves are controlled by a controller that sends signals to them. The controller compares the actual flow rate with the setpoint, which is the desired flow rate or pressure. By moving the valve, the controller will bring the flow rate or pressure to the setpoint value. Due to the design of pneumatic actuators, a control valve will fail to reach a specific position in the event of a loss of control signal. Pneumatic actuators are designed in such a way that when the control signal is lost, the valve will fail to operate to a specific position. The spring will move the valve stem out of the body as the supply air pressure is reduced. Failsafe mode refers to the position at which the control valve moves in the event of a loss of signal. There are different types of failsafe modes depending on the application of the control valve. An actuator closes a fail-closed control valve when it fails. In a fail-open control valve, the actuator opens the valve. During normal operation, the electrical or pneumatic actuator must overcome the force of the spring. An electrical or pneumatic failure of the actuator results in the spring pressure forcing the valve to open or close in the absence of electrical or pneumatic forces.

6.4.6 Flare Drums

The flare drum, also known as the flare knock-out drum, is a vessel used to remove any liquids or liquid droplets from the flare gas. Knock-out vessels are used to slow down gases and allow liquids to "fall out" of the gas stream. The purpose of this separation is to prevent the accumulation of liquid at the bottom of the flare stack that could obstruct the path of the gas. By minimizing the risk of liquid combustion at the flare tip, we can recover the fractions that can be used. The presence of liquids in the stream can extinguish flames or cause irregular combustion and smoking. Additionally, the flaring of liquids can generate a spray of burning chemicals that can reach the ground and pose a safety risk. Knock-out drums are a main component in pressure-relief systems in industries. This device is also known as a flash drum or a knock-out pot. All flare systems are designed to include a liquid knock-out drum. In the event of excessive pressure, knock-out drums and flare systems should be designed appropriately, since they can cause equipment failures that can result in economic losses for the business, environmental contamination, and health and safety risks. For knock-out drums, both horizontal and vertical designs are considered based on the operating parameters as well as other plant conditions. Horizontal drums are often more economical if a large liquid storage capacity is desired and the vapor flow is high. The majority of knock-out drums are horizontal. To allow entrained liquids to settle or drop out, a horizontal knock-out drum must have a diameter large enough to keep the vapor velocity low. The pressure drop across horizontal drums is generally the lowest of all designs. A vertical knock-out drum is typically used when the liquid load is low or limited plot space is available. Two-phase fluid enters the knock-out drum. After hitting the deflector plate, the gas and fluid partially separate first, and for fluids that remain at the bottom, the gas rises as a result of gravity.

A knock-out drum, which is used for separating gases and liquids, is shown in Figure 6.6. Using a flare main header, the relieving gases from safety relief

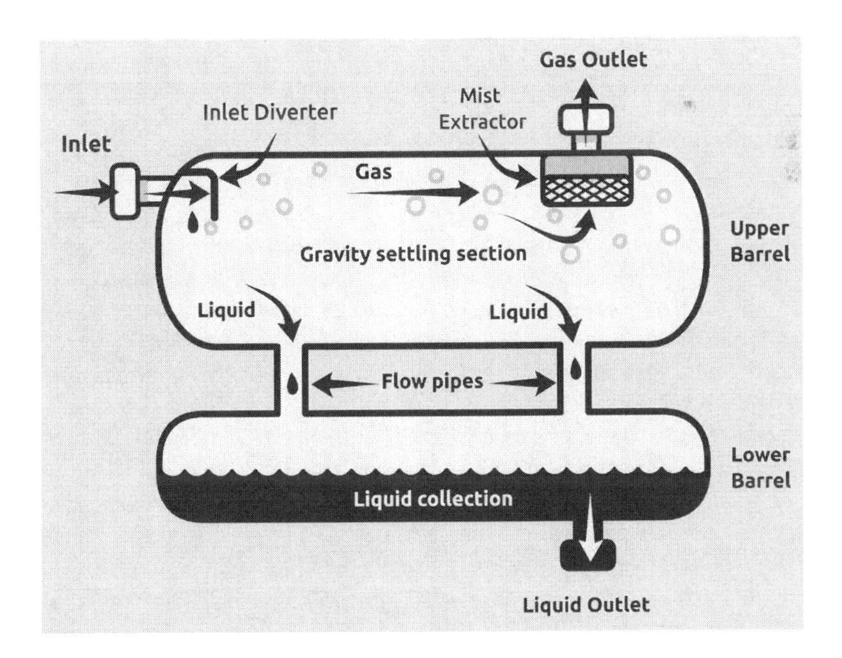

FIGURE 6.6 A knock-out drum.

(Courtesy: Shutterstock)

valves and pressure control valves are collected in a horizontal or vertical knock-out (KO) drum. The condensate that is carried along with the gases is knocked down here. The liquid level in the boot/drum is maintained at a constant level. In oil recovery facilities, the liquid is either pumped to a slop tank or reused. To prevent freezing, steam may be used for winterizing. Gas from the KO drum is then directed to an elevated flare stack. Normally, a liquid seal is maintained at the bottom of the stack. Ultimately, flare knock-out drums separate the liquid from the gas, prevent liquids from being discharged into the atmosphere, increase the life of flare tips, and allow liquids to be drained for processing or disposal.

6.4.7 Flare Manifolds

It consists of a series of lines connecting the protection components (valves) to the flare drum. It is composed of sub-manifolds and a main manifold (also called a collector or header). It is essential that all of these manifolds have a sufficient diameter so that there will be no back-pressure when several protection components are opened simultaneously. They should also be installed at a slope (e.g. 2 mm per meter) towards the flare drum to ensure that liquids carried over during burning are naturally drained.

6.4.8 Sealing Systems

Flare systems are equipped with seals that prevent flashback. In the absence of a seal, a continuous flow of gas may be bled to the flare in order to maintain a positive flow. Liquid seals and gas seals are the two main types of seals.

6.4.8.1 Liquid Seals

There are two types of liquid seals: seal drums and seal pipes. In the former, a liquid seal is used in a seal drum (see Figure 6.7) located between the KO drum and flare stack. Horizontal or vertical seal drums can be used; the choice is primarily determined by the available space. Prior to reaching the flare stack, process vent streams are usually passed through a liquid seal. In addition to being downstream of the knockout drum, the liquid seal can also be integrated within the same vessel as previously discussed. The liquid seal drum is a specially designed vessel that contains a predetermined level of water at its base. Liquid seal drums are designed to prevent flame propagation in the unlikely event of flashback. A liquid seal also serves as a large check valve that prevents gas from traveling upstream for any reason. A third function of a liquid seal is to remove liquid droplets. It is possible to design liquid seals as independent vessels or as integral components of flare risers. Once the waste/process gas enters the drum through the flare system header, it is diverted down into the water and forced to bubble through the liquid seal. As the gas travels up the flare stack and tip, it is ignited. A horizontal liquid seal drum and main parts are illustrated in Figure 6.8.

FIGURE 6.7 A liquid (water) seal drum.

(Courtesy: Shutterstock)

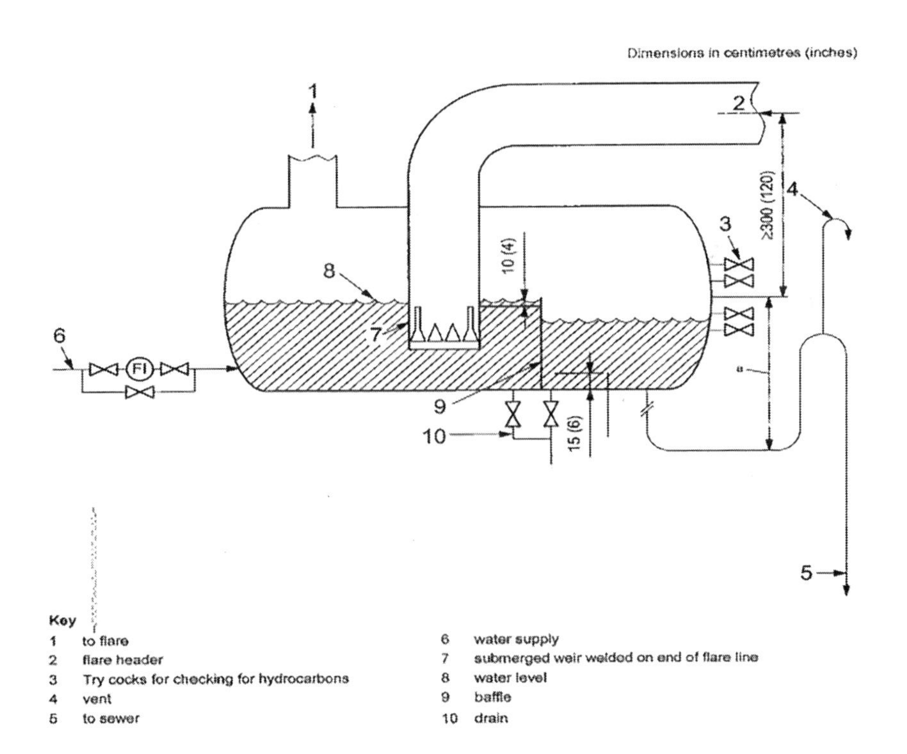

FIGURE 6.8 A horizontal liquid (water) seal drum and main parts.

For all disposal systems, flashback protection (the possibility that the flame may travel upstream into the system) must be considered due to the possibility of pressure build-up in upstream piping and vessels as a result of flashback. Flame flashbacks can be prevented by preventing air from inadvertently entering the flare system and pulling the flame front downward. A flare system is subject to explosion hazards when air is present in the system. Additionally, the liquid seal maintains a positive pressure on the upstream system and dampens any explosive shock wave in the flare stack. Sometimes, piping seals are used in place of drums at the bottom of the stack to serve as seal legs. In many cases, this is an integral part of the stack. Some liquid seals may be replaced or used in conjunction with other devices, such as flame arresters and check valves. By preventing flashback caused by low vent gas flow, purge gas also helps prevent flare stack flares. To prevent low flow flashback problems and to prevent flame instability, the total volumetric flow to the flame must be carefully controlled. It is necessary to maintain a minimum flow through the system by using purge gas, typically natural gas, N_2, or CO_2. To prevent the formation of an explosive mixture in the flare system, N_2, another inert gas, or a flammable gas must be used if air is present in the flare manifold. Purge gas injection should be made at the farthest upstream point of the flare transport piping in order to ensure a positive flow throughout all flare components. Purge gas injection is also known as a gas seal, which is discussed in greater detail in the following section.

6.4.8.2 Gas Seals

Molecular-type seals have been developed more recently to prevent flashbacks in flare systems. The purge gas must have a molecular weight of 28 or less (for example, N_2, CH_4, or natural gas). As a result of the buoyancy of the purge gas, a zone with a greater pressure than atmospheric pressure is created. There is a molecular seal located at the top of the flare stack immediately below the burner tip. Due to the high pressure in the stack, ambient air cannot enter. A gas seal is shown in Figure 6.9.

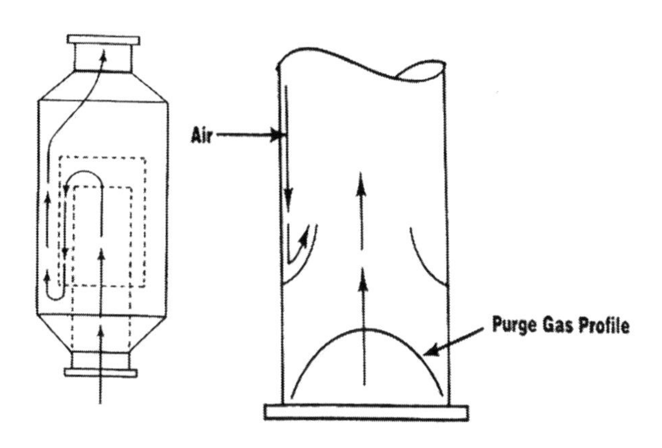

FIGURE 6.9 A gas seal.

6.4.9 Flare Stacks

Flare stacks are vertical-axis towers that burn waste or contaminated products using an open flare. In refineries, chemical plants, and landfills, flare stacks are used to burn off unusable waste gases or flammable gases and liquids released by pressure relief valves during unplanned overpressuring. An example of a flare stack in a refinery can be seen in Figure 6.10.

A variety of flare stacks are available, including conventional stacks, sonic stacks, low flares with combustion chambers, and cold flares or vents. As a rule, the conventional stack is always installed vertically, and the gas velocity is limited to Mach 0.5/0.6 for discontinuous flows (emergency shutdown) and Mach 0.3 for continuous flows. A flare stack must be able to function under all atmospheric conditions and must have a reliable ignition system. A flare stack's height or the reach of a flare boom is determined by the amount of thermal radiation that is permissible or tolerable for equipment or personnel. Sonic flares are a type of high-pressure flare tip that is unique. Sonic tips are designed to discharge the flare gas at sonic velocities. The purpose of a sonic flare is to eliminate smoke, reduce flame radiation, and shorten the flame length by using the flare gas pressure. As a result of lower stack heights and a smaller flare header size, sonic flares can reduce capital costs. A low flare with a combustion chamber consists of a chimney that houses a forced-draught burner. They are installed on land when the environmental regulations do not allow for a visible flame or when there is insufficient space to install another type of flare. Furthermore, they can also be installed offshore on a ship when it is not possible to install another type of flare. Cold flares are similar to other types of flares, except that the gas is released into

FIGURE 6.10 A flare stack in a refinery.

(Courtesy: Shutterstock)

the atmosphere rather than being burned. Cold flare height can only be determined by calculating the dispersion of gases in the atmosphere.

6.4.10 Flare Ignition Systems

The ignition system is one of the most important features of any flare system. Pilot, ignition, and flame detection systems are integral parts of any flare system supply. Generally, pilot burners are recommended for igniting flares, with the exception of a few flare applications. Environmental Protection Agency (EPA) regulations require flares to have a continuous flame. This is achieved through continuous pilot burners that provide a reliable and stable ignition. Pilot burners are located around the outer perimeter of the flare tip. Pilot burners are essentially premixed burners designed to provide a stable flame to ignite waste gas exiting the flare tip. They consist of three main components, the Venturi or mixer, gas jet or orifice, and burner nozzle. It is necessary to use an ignition system in order to light the pilot burner and, in turn, the flare tip. These pilots are equipped with wind shields so that the most severe wind cannot blow them out.

Flame front generators (sometimes referred to as "fire ball" ignition types) are the most common type of flame generator. By means of control valves, compressed air (generally instrument air or plant air) and fuel gas are metered into a mixing chamber located on a panel at grade. The sparking/ignition device is located downstream of the mixing chamber. The flame front line is connected to the pilot burner nozzle by a dedicated line. In the combustion chamber, the gas/air mixture is ignited and a flame front fire ball is generated. After traveling along the flame front line, the flame front ignites the pilot burner. It is an advantage of the compressed air flame front generator that the flow controls and the sparking device are at grade and can be serviced during operation.

6.4.11 Flare Burners or Flare Tips

An area of the flare where fuel and air are mixed at the speeds, turbulence, and concentration necessary to ignite and maintain a stable combustion process. Flare tips are large burners or a combination of smaller burners installed at the end of flare risers for the purpose of safely igniting and burning flare gas discharges. In recent decades, flare tip designs have evolved significantly. Although many suppliers offer proprietary flare technologies, the basic types of flare tip designs can generally be categorized as follows:

A ***single-point flare tip*** is typically positioned vertically and attached to an elevated flare riser or header. In offshore operations, single-point flares are sometimes positioned at an angle on the end of a flare boom in order to direct the flame away from the operating platform. There are two types of flares: subsonic pipe flares and sonic single-point flares. A flare boom is a structure that is typically attached to an oil rig. It allows the safe burning of gas that cannot be collected safely during the extraction of oil from an offshore oil rig. The end is extended away from the

> main oil platform in order to keep the burning flame as far as it is safe from the platform, so as not to endanger people, equipment, or the platform. The system may be onshore, in which case it is commonly referred to as a flare stack or gas flare.

In general, ***multi-point flare tips*** are sonic flares and offer improved performance with regard to radiation, noise, and smokeless operation. This is achieved by routing the gas to a number of smaller diameter burner nozzles. Aerating the flared gas in the combustion zone is facilitated by breaking up the gas flow into smaller streams. This allows greater air contact with hydrocarbon gas in the combustion zone. Radiation from a multipoint flare can be reduced by up to 40% compared to a single-point tip of the same size. Similar to single-point flares, these flare tips generally have a flanged connection at the top of an elevated flare riser/header. The comparison between a single-point flare tip and a multi-point flare tip is shown in Figure 6.11.

An ***air-assisted flare tip*** provides clean combustion by injecting primary air into the flame's base through an electrically powered fan/blower located at grade. The combustion efficiency of flared gas can be increased by installing an air blower, which will reduce smoke formation during the combustion process. The air-assisted flare tip incorporates a design that divides the waste gas stream into a number of smaller streams at the top of the flare in order to increase the gas/air contact surface area and ensure that the gas is better mixed with the air. Flows of forced air from the blowers and gas from the flare header are routed separately from the flare stack's base to its top.

FIGURE 6.11 Comparison between a single and multi-point flare tip.

In ***steam-assisted flares***, the burner tips are elevated above ground level for safety reasons and burn the vented gas in what is essentially a diffusion flame. The majority of flares installed in refineries and chemical plants are of this type, and they constitute the predominant flare type. To ensure an adequate air supply and good mixing, this flare system injects steam into the combustion zone to promote turbulence for mixing and to induce air into the flame to ensure an adequate air supply. A steam-assisted flare often achieves smokeless combustion more effectively than an air-assisted flare because high-pressure steam can supply more momentum, which enhances ambient air entrainment and air-fuel mixing. For flares of similar size, where steam is available, steam-assisted flares have a lower capital cost and a wider operating range than air-assist flares.

It is important to note that the *non-assisted flare* consists of only a flare tip without any additional means of enhancing the mixing of air into its flame. In general, its application is limited to gas streams with low heat content and a low carbon to hydrogen ratio that burn readily without generating smoke. For complete combustion of these streams, less air is required.

British Petroleum's flare department developed *Coanda flare* tips in the 1960s for use on their North Sea production platforms. Following the success of the design, BP established a subsidiary company to market the product worldwide. The Coanda effect is used to entrain large volumes of air into the gas stream, resulting in a single clean-burning compact flame. The Coanda effect refers to the tendency of fluid discharging from a nozzle or slot to adhere to an adjacent convex surface, entertaining the surrounding fluid.

6.4.12 Instrumentation and Control

A typical flare system instrumentation and control system consists of the following components:

1. In order to achieve smokeless burning, a suitable control system is provided for regulating steam injection into flare tips. In most flare headers, a flow sensor is provided. Alternatively, a flame monitoring device measures the light output of the flame, which is used to set the steam flow to maintain smokeless operation.
2. Pilots are provided with thermocouples in the control room which are connected to an alarm.
3. As part of the flare system, an oxygen analyzer with an alarm is typically provided to indicate the pressure of the air or oxygen.
4. The flare knockout drum requires continuous, accurate level monitoring to prevent liquids from reaching the flare and creating a fire hazard. It is essential to track and continuously report to the operator the total process level over the span of the drum in order to ensure the safety of the process.
5. There is a flare video monitor in the control room that allows smokeless operation to be observed and abnormal flare header releases to be identified.

6.5 FLARE SYSTEM WORKING PRINCIPLES

By burning hydrocarbons and creating safe carbon dioxide and water vapor, a flare system collects hydrocarbon releases from relief valves, blowdown valves, pressure control valves, and manual vents. Waste gas, also known as unused gas, is collected and transported through valves and a network of pipes. If the pressure increases too sharply and/or becomes uncontrollable, the equipment's safety valves open to protect the vessel. Each drum, column, or capacity that operates under hydrocarbon pressure is linked to the flare network by means of one or many valves and/or various pressure control valves (PCVs) and blow down valves (BDVs). The liquid is removed from the gas using a knock-out drum, which then goes to the drain. The pipes contain a sealed system designed to prevent explosive mixtures from occurring. The gas flows up the header, an elevated stack that keeps the flare away from personnel and facilities, causing a chain reaction that results in combustion at the top due to the pressure created upstream. By means of a series of small continuous flames, the pilot light ignites the gas. The steam line at the tip of the header injects steam into the flame, thereby reducing soot and smoke emissions. When too much steam is added, a condition known as "over-steaming" can occur, resulting in reduced combustion efficiency and higher emissions.

6.6 DESIGN CONSIDERATIONS FOR FLARE SYSTEMS

It is important to consider the gas sources when designing a flare, as the ultimate flare system will be determined by these sources. Various pressure protection devices such as relief valves and rupture discs are typically used to automatically release gases from process equipment. Design consideration for flare systems must include the following:

1. Heat intensity radiated from a flare that may pose a potential danger to personnel and equipment nearby.
2. The presence of hydrogen sulfide in the vapors (gases) produces sulfur dioxide and as such can cause damage to vegetation at ground level if sufficiently high concentrations are present.
3. As a result of inefficient combustion air distribution, hydrocarbon vapor emitting from a plain open-ended tip tends to burn with an unstable flame and produce large quantities of smoke.
4. A continuous source of ignition is required.
5. In a short period of time, vapor (gas) flows can vary from zero to very high rates and vice versa.
6. Vapor compositions can vary due to emissions from different sections of a plant.
7. In essence, the blowdown header system is a hydrocarbon vapor pipe open to the atmosphere at the flare point.

8. The expansion of high-pressure hydrocarbon vapors through pressure relief devices results in partial liquefaction and cooling of the vapors within the blowdown header system. A knock-out device is necessary to eliminate liquid carry-over into elevated flares.
9. Typical flare design considerations are flow rate, temperature, header sizing, back-pressure, smokeless operation, flash back protection, ignition system, location, and environmental considerations.

6.7 NOISE POLLUTION

Petrochemical plants have accepted noise pollution caused by flares for too long as an inevitable byproduct of the flaring process. The major source of noise from flares is usually located at the flare tip itself. This is especially true when the flare tip is used for smokeless flaring of hydrocarbon gases through steam injection. Basically, noise is produced by two factors, steam energy losses at high pressure steam injectors and unsteadiness in the combustion process. Generally, closed ground flares produce less noise than elevated flares. As a result of being enclosed in a box, the flame is protected from wind effects, and the heat re-radiated from the refractory walls stabilizes the combustion process and reduces the random characteristics. As a result of the walls themselves absorbing some of the sound energy, some sound will be absorbed. The development of sophisticated flare tips has greatly reduced the level of noise pollution. The combustion efficiency of some designs has been greatly improved by mixing air with gas before combustion takes place. Before the gases exit the flare tip, steam is also pre-mixed with air and gas. Consequently, some of the turbulent noise energy is shielded by the tip.

6.8 SMOKELESS OPERATION

When there are burning carbon particles present in a flame, it is referred to as being luminous. Smoke is formed when these particles cool. Flares are normally designed to prevent smoke in three different ways: by adding steam; to ensure efficient combustion, fuel and air should be premixed before combustion in order to provide sufficient oxygen; and a number of small burners are used to distribute the flow of raw gases. For economy and superior performance, steam is most commonly used to produce a smokeless flare. Prior to entering the combustion zone of a flare, raw gas is preheated using steam addition. It is possible for the hydrocarbons to crack if the temperature is high enough. As a result, hydrogen and carbon are produced in free form. When the cracked hydrocarbons travel to the combustion zone, hydrogen reacts much faster than carbon. It is inevitable that the carbon particles will cool down and form smoke if they are not burned away. Therefore, in order to prevent smoke, either the hydrogen atom concentration must be reduced in order to ensure uniform combustion of both hydrogen and carbon, or sufficient oxygen must be provided. Therefore, the formation of black smoke can occur when the flame does not have enough oxygen to completely burn the waste gas. The flare may smoke if insufficient amounts of steam or air are injected into the flare tip.

6.9 FLARE TERMINOLOGY

For the purpose of improving the clarity of this chapter, some of the essential flare terms are defined in this section.

Generally, *assist air* refers to all air that is deliberately introduced into a flare tip via nozzles or other hardware conveyances in order to promote turbulence for mixing or inducing air into the flame, and/or protect the flare tip. Air provided as assistance does not include ambient air.

It is the purpose of *assist steam* to introduce steam into a flare tip prior to or at the flare tip using nozzles or other hardware conveyances in order to protect the flare tip design or promote turbulence so that air can be mixed into the flame.

An *auxiliary fuel* is any gas that is introduced to the flare in order to increase the heat content of the combustion zone gas.

Combustion zone gas refers to all gases and vapors found just after a flare tip is lit. All flare vent gas, all assist steam, and any assist air, if any, that is intentionally introduced to the flare vent gas or center steam prior to the flare tip are included in this gas.

A *flare purge gas* is introduced between a flare header's water seal and the flare tip in order to prevent oxygen from infiltrating into the flare tip (backflow). *Flare sweep gas* performs the function of flare purge gas in flares without water seals.

Flare sweep gas refers to gas deliberately introduced into the flare header system in order to maintain a constant flow of gas through the flare header in order to prevent oxygen from building up in the flare header and, for flares without flare gas recovery systems, to prevent oxygen from infiltrating the flare tip.

Flare vent gas refers to all gases found just prior to the flare tip. There are several gases included in this category, including flare waste gas, flare sweep gas that is not recovered, flare purge gas, and auxiliary fuel, but it does not include pilot gas, assist steam, or assist air.

The term *flare waste gas* refers to the gas generated during facility operations that is directed into a flare for the purpose of disposal.

Pilot gas is gas introduced into the flare tip that provides a flame for igniting the flare vent gas.

QUESTIONS AND ANSWERS

1. Identify the incorrect statements about gas flaring in the oil and gas industry.
   A. As part of the oil extraction process, gas flares are used to burn natural gas.
   B. The purpose of flaring is not to improve safety but rather to meet legal or economic requirements.
   C. The liquids are separated from the gases in knock-out drums and are sent to the flare stack for combustion.
   D. As part of the flare system, thermal relief valves can be called thermal safety valves, temperature relief valves, or thermal expansion relief valves.

Answer) The statements in options A and D are correct. Option B is incorrect because gas flaring is primarily motivated by safety concerns. Additionally, option C is incorrect because the gases that are separated in knock-out drums are sent to the flare stack as well.

2. In the flare system, what type of valves are used?
   A. Globe and ball valves
   B. Pressure safety and control valves
   C. Butterfly valves
   D. Modular valves

   Answer) Option B is the correct answer.

3. In terms of the main purpose of liquid seal drums, which of the following statements is correct?
   A. The main purpose of the liquid seal drum is to prevent flame propagation. It quenches the flame with a barrier of water, as well as providing other intrinsic design benefits.
   B. By acting as a large check valve, the liquid seal prevents any gas from traveling upstream.
   C. Droplets of liquid are disturbed by it.
   D. All three choices are correct.

   Answer) Option D is the correct answer.

4. What component or facility of a flare system is responsible for separating liquids from gases?
   A. Knock-out drum
   B. Liquid seal
   C. Flare stack
   D. Separator

   Answer) Option A is the correct answer.

5. What is a vertical-axis tower of flare that burns waste or contaminated products using an open flare?
   A. Flare boom
   B. Flare stack
   C. Flare tip
   D. All three choices are incorrect

   Answer) Option B is the correct answer.

6. Determine which statement about ground flares and elevated flares is incorrect.
   A. The selection of a flare is influenced by a number of factors, including the availability of space; flare gas characteristics (i.e. composition, quantity, and pressure level); and economics, including the

initial investment and operating costs, as well as concerns regarding public relations with the surrounding community.
B. Ground flares have the disadvantage of having to be isolated from the remainder of the plant and process lines, thereby requiring a considerable amount of space and long interconnecting piping.
C. Maintenance of elevated flares is more difficult than ground flares.
D. All three choices are correct.

Answer) Option D is the correct answer.

7. For flare systems, what type of instrumentation and control is used?
A. Thermocouple for knock-out drum
B. Level measurement for the flare tip
C. An oxygen analyzer with an alarm is typically provided
D. Video monitoring on the flare

Answer) Option A is incorrect since thermocouples are used for the pilots rather than knock-out drums. In the case of knock-out drums, level measurement is more important. As a flare tip does not require level measurement, option B is incorrect. C and D are both valid options.

8. What kind of flare stack results in a reduction in smoke and radiated heat?
A. Conventional flare stack
B. Sonic flare stack
C. Cold vent
D. All three choices are correct.

Answer) Option B is correct.

9. In order to prevent smoke from flares, which approach is more effective and popular?
A. Adding steam to the flare.
B. Before combustion, a premix of fuel and air is prepared in order to provide sufficient oxygen for efficient combustion.
C. Using a number of small burners to distribute the flow of raw gases.
D. All three choices are incorrect.

Answer) Option A is correct.

10. Which flare component is used prevent flame propagation and flashback?
A. Safety valves
B. Check valves
C. Liquid or gas seals
D. Knock-out drum

Answer) Option C is the correct answer.

FURTHER READING

1. Argo Flare. (2023). *Flare types*. [online] available at: www.argoflares.com/research/introduction/flare-types/ [access date: 26th July, 2023]
2. Argo Flare. (2023). *Pilot burners, ignition systems and flame detection*. [online] available at: https://argoflares.com/research/introduction/pilot-burners-ignition-systems/ [access date: 26th July, 2023]
3. Bader, A., Baukal, C., & Bussman, W. (2011). Selecting the proper flare systems. *Chemical Engineering Progress*, 45–50.
4. Emam, E. A. (2015). Gas flaring in industry: An overview. *Petroleum & Coal*, 57(5).
5. The World Bank. (2023). *Global gas flaring reduction partnership*. [online] available at: www.worldbank.org/en/programs/gasflaringreduction/gas-flaring-explained [access date: 23rd July, 2023]
6. Sutton, I. (2015). *Plant design and operations*. Oxford, UK: Elsevier. ISBN: 978-0-323-29964-0
7. Total. (2007). *Process drains and flares: Training manual, revision 1*. [online] available at: https://www.scribd.com/document/239515436/Flares
8. Vallavanatt, R., & Self, F. (2021). Gas flare systems—last line defense. *Process Safety Progress*, 40(3), 132–148.

7 Electrical Components

7.1 INTRODUCTION

A vital component of any operation is the electrical system. Pumps and other equipment used in the production and treatment of oil can be powered by electricity. As well as powering lights, tools, and even some conveniences, such as a radio, the system will also provide power for other appliances. A well-designed electrical system is critical to ensuring continuous production from an oil field. When designing a power system, the engineer is required to estimate the normal operating load of the plant. In addition, they are interested in knowing how much additional margin should be included in the final design. There are no strict rules for estimating loads, and various basic questions need to be addressed at the outset of a project, such as: Is the plant a new, greenfield plant? What is the expected lifespan of the plant, for example, 10, 20, 30 years? Is the plant old and undergoing an expansion? The objective of this practical project is to investigate the design of the power supply in an oil and gas refinery. There are usually two main stages to the design process in these types of projects: the basic design stage and the detail design stage. An analysis of the location and process of the refinery is conducted in the basic design stage, and a rough estimate of the types of loads and demand is derived as a result. In the next step, various kinds of networks and supplies are surveyed (based on technical and economic conditions) and the best choice is selected, as well as basic calculations, drawings, and specifications. Identifying the areas of engineering and design where interfaces are necessary should be carried out at the earliest practical stage of a project. At regular intervals, an efficient system of communication and information exchange should be established and implemented. Electrical engineers should be informed on matters relating to production processes and supporting utilities by the process engineers. Mechanical engineers will normally be required to provide information regarding power consumption data for rotating machines, such as pumps, compressors, fans, conveyors, and cranes. Generally, the process and instrument engineers are responsible for developing the operating and control philosophies for individual equipment as well as overall schemes. Following this, the electrical engineer should interface to enable the following to be understood.

Industrial plants, including oil and gas and petrochemical plants, are primarily designed to produce in a consistent and economical manner. Availability and continuity of electrical service play an important role in determining the ability to produce. A service interruption can usually be evaluated directly in terms of lost production in many cases. It is possible that the costs associated with this interruption may exceed the costs associated with the physical damage to the electric equipment. Consequently, it is of utmost importance that the electrical system be

DOI: 10.1201/9781003465881-7

designed to provide continuous, reliable, and hopefully undetected service. Due to the fact that no two plants have the same requirements, a standard electrical system cannot be applied universally. Therefore, it is essential to analyze the specific requirements of each process system and then design an electrical system that will most effectively meet those requirements. An oil field's electrical system includes power generation, power distribution, electric motors, system protection, and electrical grounding. There are two options for generating electricity on site: either generating it on site or purchasing it from a local utility.

7.2 POWER GENERATION SYSTEMS

Power generation in remote oil and gas facilities is typically provided by diesel or gas-fueled reciprocating engine generator packages or gas turbine generators. An individual generator rated above 1000 kW that is intended for use in the oil industry is usually driven by a gas turbine (also known as the prime mover). Under 1000 kW, diesel engines are usually preferred, typically because they are emergency generators that use diesel oil fuel. Many offshore facilities with power demands below 1000 MW also use gas-fueled reciprocating engines as the prime mover. By using one or more reciprocating pistons, a reciprocating engine, also known as a piston engine, converts high temperatures and high pressure into rotation. Figure 7.1 shows a diesel generator. As the heart of a power plant, a gas turbine converts natural gas or other liquid fuels into mechanical energy. This energy then drives a generator that produces the electrical energy that moves along power lines to homes and businesses, as shown in Figure 7.2. Electricity has been generated by gas turbines since 1939. The gas turbine is one of the most

FIGURE 7.1 A diesel generator.

(Courtesy: Shutterstock)

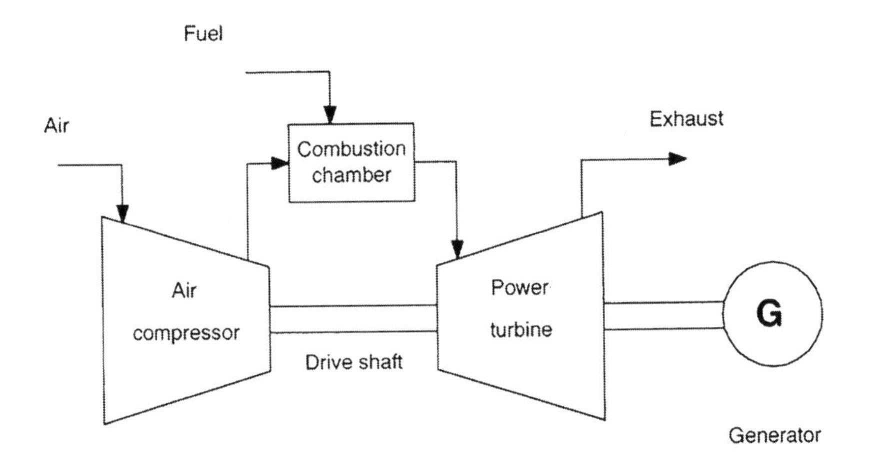

FIGURE 7.2 A gas turbine and generator.

widely used technologies for generating power today. The gas turbine is a type of internal combustion (IC) engine that produces power by burning an air-fuel mixture to produce hot gases that spin a turbine. The name "gas turbine" is derived from the hot gas produced during the combustion of fuel rather than from the fuel itself. In gas turbines, combustion occurs continuously as opposed to intermittently in reciprocating engines. Diesel or natural gas may be used as a fuel by the engines or turbines. There are some units that are dual fueled, utilizing both natural gas and diesel fuel. For most applications, natural gas-fueled prime movers are most suitable for power generation. Whenever natural gas is unavailable and emergency power is required, diesel is used.

The power supply has a greater impact on system reliability than any other component. Power, regardless of whether it is supplied by a utility company or generated on site, must be reliable. In addition to causing costly production downtime and equipment damage, electrical failures can also lead to employee injury. Dips in voltage are also particularly problematic when computers or high intensity discharge lighting (e.g. mercury vapor and high-pressure sodium) are being used. To determine the quality of power provided by a utility, it is important to investigate its outage history and past outages. In the event that a facility elects to generate its own power, quality and reliability of the generated power as well as economic considerations must be taken into account. There are many economic factors to consider when determining whether to produce power or to purchase it from a local utility. It is important to remember that many operations, including refineries, require dual sources of power. An evaluation should be conducted to determine whether it is feasible (primarily economically) for a facility to generate its own electricity. If the facility does not have waste gases, waste heat, or other fuels available, it should be determined whether fuel is available and how much it costs.

7.2.1 Electrical Generators

A generator (see Figure 7.3) is a device that converts motion-based energy (potential and kinetic energy) or fuel-based energy (chemical energy) into electric power for use in an external circuit. In most cases, electric generators are powered by rotating shafts generated by engines or turbines. There are many generator sets that are relatively small in size, typically ranging from several kilowatts to several megawatts. It is often necessary for these units to come online and operate quickly. For them to be effective, they must be capable of running for an extended period of time. For the majority of these applications, an internal combustion engine is an excellent choice as the prime mover. It is also possible to use turbines. There are a variety of fuels that are used in engines, including diesel, natural gas, digester gas, landfill gas, propane, biodiesel, crude oil, steam, and others. Several factors should be considered when selecting a generator, including cost, location (onshore or offshore), fuel source, power demand, and maintenance.

There are two types of generators: synchronous and asynchronous. Generators that operate asynchronously are also known as induction generators. Through the process of electromagnetic induction, a synchronous generator converts mechanical power into alternating current (AC) electric power. Synchronous generators are also known as alternators or AC generators. The term "alternator" refers to the device that produces AC power. Due to the fact that it must be driven at synchronous speed in order to produce AC power with the desired frequency, it is called a synchronous generator. Synchronous generators consist of two main components; a stator is the stationary component of a synchronous generator. It contains the armature winding that generates the voltage. The stator provides the synchronous generator's output. As for the second component of the alternator, the rotor is the

FIGURE 7.3 An electrical generator.

(Courtesy: Shutterstock)

rotating part. The main flux of the field is produced by the rotor. The working principle of an alternator or synchronous generator is based on electromagnetic induction, that is, when the flux surrounding a conductor changes, an electromotive force is generated in the conductor. The alternator armature winding will generate voltage when it is subjected to a rotating magnetic field. The induction generator or asynchronous generator is a type of alternating current electrical generator that produces electricity by using the principles of induction motors. By mechanically turning their rotors faster than synchronous speed, induction generators produce electricity. There are certain advantages to using an induction generator over a synchronous generator. As an example, voltage and frequency are controlled by the utility; therefore, voltage and frequency regulators are not needed. Furthermore, the generator construction is highly reliable and requires little maintenance. A minimum number of protective relays and controls is also required. There are some major disadvantages to this type of generator, including the fact that it requires variables from the system and cannot normally be used as a standby or emergency generator. The most common type of generator is a synchronous generator.

There are primarily three types of generator systems. First, there is the single generator that operates independently of the electric utility power grid. Typically, this is referred to as an emergency standby generator. As shown in Figure 7.4, there is a single standby generator, a utility source, and a transfer switch. The load is either supplied by the utility or by the generator in this case. There is never a continuous connection between the generator and the utility. There are few requirements for protection and control in this simple radial system. Additionally, it has the least impact on the entire electric power distribution system. Although this type of generator system improves overall electrical reliability, it does not provide the redundancy that some facilities require in the event of a generator failure

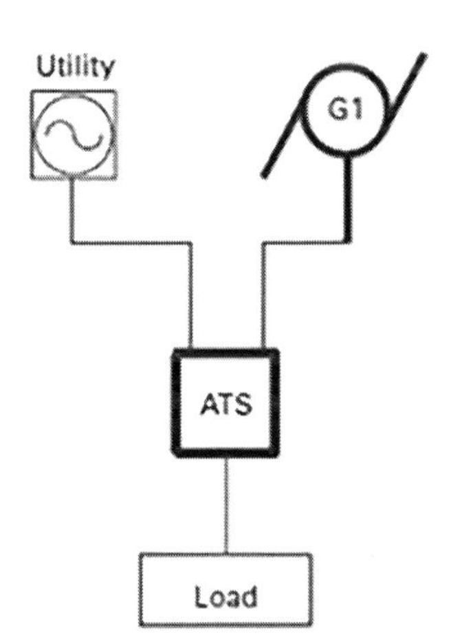

FIGURE 7.4 An emergency standby generator system.

FIGURE 7.5 An electrical motor under inspection.

(Courtesy: Shutterstock)

or maintenance period. There is a second type of generator system that consists of multiple isolated standby generators. If one generator fails to start or is out for maintenance, it will not affect the load. The third type of system is either one with a single or multiple generators that operate in parallel with the utility system.

7.2.2 Electrical Motors

Motors are electrical machines that convert electrical energy into mechanical energy. An electric motor generates torque by interacting with a magnetic field and an electric current in a wire winding to produce force. In the oil and gas industry, electric motors provide reliable and consistent power for drilling rig systems and equipment. As a result of the power generated by these motors, crude oil, petroleum, and natural gas are extracted, processed, stored, and transported. Three-phase AC induction motors are the most common industrial motors due to their reliability and low cost. As a result of the electric current flowing in the stator winding, a rotating magnetic field is generated, which induces an electric current in the rotor. Most minor processes at a refinery are powered by electric motors, which account for over 80% of the electricity consumed. Approximately 60% of the motors used are used to drive pumps, 15% for air compressors, 9% for fans, and 16% for other purposes. An electrical motor is shown in Figure 7.5 during an inspection.

7.3 POWER DISTRIBUTION SYSTEMS

Electric power distribution is the final stage in the delivery of electricity. Electricity is carried from the transmission system to individual consumers. Electricity is dominant since it is relatively easier to transmit and distribute than other forms of energy, such as mechanical energy. It is necessary to bring electricity to an oil field at higher voltages in order to reduce power losses. The primary distribution system is a higher-voltage distribution system. Distribution substations are located near or within a city/town/village/industrial area. Power is received from a transmission network. Using a step-down transformer, the high voltage from the transmission line is stepped down to the primary distribution level voltage. The secondary electrical distribution system consists of devices that operate at the same voltage as the motors, such as the transformer at the end of the primary system, the cables, and the disconnect switches. The main components of the power distribution system are explained as follows.

7.3.1 Distribution Substations

Power is transferred from the transmission system to the distribution system of an area via a distribution substation (see Figure 7.6). As it is not economically viable to connect electricity consumers directly to the main transmission network, unless they consume large amounts of energy, the distribution station reduces the voltage to levels suitable for local distribution.

7.3.2 Supply Lines or Feeders

A distribution feeder (supply line) is a dedicated line of cables that transport power from the distribution substations to the premises of the end users. Distribution substations are connected to sub-transmission systems through at least one supply line, known as a primary feeder, which serves a large number of premises. In general, a distribution substation is supplied by two or more supply lines in order to increase reliability of the power supply in the event that one supply line

FIGURE 7.6 A distribution substation.

(Courtesy: Shutterstock)

is disconnected. High-voltage disconnecting switches are used to isolate supply lines from the substation so that maintenance or repair work can be performed. To ensure that electricity is delivered safely and reliably, electricians use a variety of feeders. To connect the generating station or substation with the main distribution points, high-voltage transmission lines are used. Power is then transmitted from these main points to homes and businesses via medium-voltage distribution lines.

7.3.3 Transformers

Transformers "step down" supply line voltage to distribution-level voltage. The distribution transformer, also known as the service transformer, is responsible for performing the final transformation in the electric power distribution system. The transformer is essentially a step-down three-phase transformer. The voltage is stepped down by the distribution transformer. High-voltage transformers are shown in Figure 7.7.

7.3.4 Busbars

Busbars (also called bus bars) are metallic strips or bars used for high-current power distribution in switchgear, panel boards, and busways. A busbar or bus can be found throughout the entire power system, from the generator to the industrial plant to the electrical distribution board. A busbar is used to carry large currents and to distribute current to multiple circuits within switchgear or equipment. Busbars are typically made of copper or aluminum and are rigid and flat—they are wider than cables but are approximately 70% shorter. Additionally, they are capable of carrying a greater amount of current than cables with a similar cross-sectional area. It is important to note that bus bars are arranged in a variety

FIGURE 7.7 A voltage transformer.

(Courtesy: Shutterstock)

FIGURE 7.8 Copper busbar arrangement.

(Courtesy: Shutterstock)

of configurations; the purpose of any particular arrangement of bus bars is to achieve adequate operating flexibility, sufficient reliability, and minimum cost. Figure 7.8 shows an arrangement of metallic busbars made of copper.

7.3.5 Switchgears

A switchgear (Figure 7.9) is a general term for primary switching and interrupting devices and their control and regulating equipment. Among the components of power switchgears are breakers, disconnect switches, main bus conductors,

FIGURE 7.9 Electrical switchgear.

(Courtesy: Shutterstock)

interconnecting wiring, support structures with insulators, enclosures, and secondary devices for monitoring and controlling the power supply. Switchgears can be of outdoor or indoor types or a combination of both.

7.3.6 Switching Components

In order to connect or disconnect elements of the power system, switching apparatus or components are required. Among the switching apparatus are switches, fuses, circuit breakers, and service protectors.

7.3.6.1 Switches

Switches are electrical components used in electrical engineering to either disconnect or connect the conductors in a circuit, interrupting the electric current or diverting it from one conductor to another. A switch is used to isolate a load, interrupt a load, or transfer the service between different sources of supply. An isolation switch provides a visible disconnect so that isolated equipment can be accessed safely. Switches are usually electromechanical devices consisting of one or more sets of movable electrical contacts connected to external circuits. If two contacts are touching, current can flow between them, whereas when the contacts are separated, no current can flow between them. An individual may manipulate a switch directly to send a control signal to a system, such as a computer keyboard button, or to control the flow of power in a circuit, such as a light switch.

7.3.6.2 Fuses

The fuse is an electrical safety device used to protect electrical circuits from overcurrents in electronics and electrical engineering. In this device, the essential component is a metal wire or strip that melts when too much current flows through it, thus stopping or interrupting the flow of electricity. The device serves as a sacrifice. Fuses are available in a variety of voltage, current, interrupting, and current-limiting ratings, as well as for indoor and outdoor use. In the event that a fuse has operated, it has created an open circuit, which requires replacement or rewiring, depending on the type of fuse. Since the beginning of electrical engineering, fuses have been used as essential safety devices. A fire accident is shown in Figure 7.10, which was caused by an electrical fault and old and unsafe fuses.

7.3.6.3 Circuit Breakers

Circuit breakers are electrical safety devices designed to protect electrical circuits from damage caused by overcurrents. It is designed to interrupt current flow so that equipment can be protected and fire risks can be minimized. In contrast to a fuse, which can only be operated once and must be replaced, a circuit breaker can be reset (either manually or automatically) in order to resume normal operation. As shown in Figure 7.11, an automatic circuit breaker is provided. A circuit breaker is a device that opens and closes a circuit either automatically or manually. It is imperative that an automatic circuit breaker is capable of opening a circuit automatically when subjected to a predetermined overload of current within

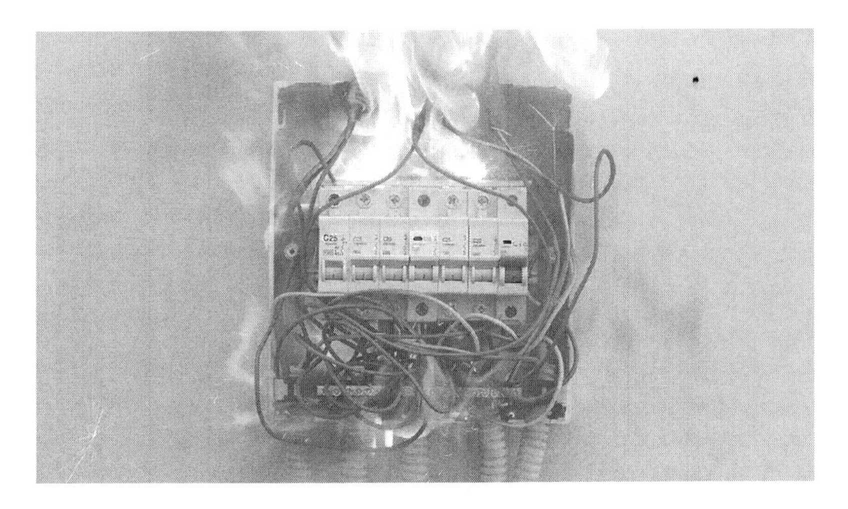

FIGURE 7.10 An electrical fault and old and unsafe fuses caused a fire accident.

(Courtesy: Shutterstock)

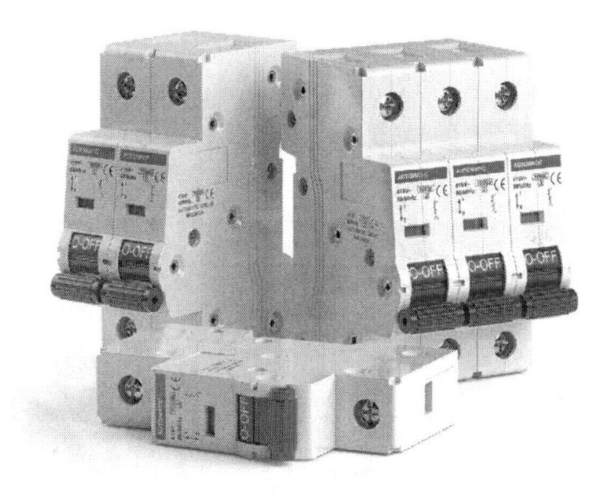

FIGURE 7.11 An automatic circuit breaker.

(Courtesy: Shutterstock)

the limits of its rating. Generally, circuit breakers must be operated infrequently, although some classes of circuit breakers can be operated more frequently.

7.3.6.4 Surge Voltage Protection

A transient overvoltage occurs as a result of the natural and inherent characteristics of power systems. In some cases, overvoltage are caused by lightning or by sudden changes in system conditions (such as switching operations, faults, etc.).

7.4 ELECTRICAL PROTECTION SYSTEMS

Power systems are affected by disturbances that may be caused by natural phenomena such as lightning, wind, trees, or animals, or by human errors or accidents. There is a possibility that these disturbances may lead to abnormal system conditions such as short circuits, overloads, and open circuits. The greatest concern is short circuits, which are also referred to as faults. They can cause damage to equipment or system elements as well as other operating problems such as voltage drops, frequency reduction, loss of synchronization, and system malfunction. Therefore, it is necessary to develop a device or group of devices capable of detecting a disturbance and acting automatically so as to prevent any adverse effects on the system element or the operator. The protection system provides this capability. Protection systems automatically disconnect faulty system elements when short-circuit currents are high enough to constitute a direct danger to the element or to the system as a whole. The protection system will initiate an alarm when the fault results in overloads or short circuit currents that do not present an immediate threat. In this section, the main electrical protection system components are relays, circuit breakers, tripping components, current transformers, voltage transformers, linear couplers, and transducers.

7.4.1 RELAYS

A relay (see Figure 7.12) is an electrically operated switch that opens and closes circuits by receiving electrical signals from outside sources. Protective relays

FIGURE 7.12 An electrical relay.

(Courtesy: Shutterstock)

serve primarily to isolate faulty sections. As a result, relays may be designed to detect and measure abnormal conditions and close the contacts in the tripping circuit. The relay may also be designed to initiate or permit the opening of various interrupting devices or sound an alarm. Based on their construction, protective relays can be classified into two main categories: electromechanical and solid-state relays. As a result of the signal provided by the transducer, an electromechanical relay develops electromagnetic force or torque, which is then used to physically open, or close, a series of contacts to permit or initiate the trip of circuit breakers or the activation of an alarm. A solid-state, or static, relay is energized by the same signal as an electromechanical relay. However, the relay contacts are not physically opened or closed. A solid-state device is used to simulate the switching of relay contacts by changing its status from conducting (closed position) to non-conducting (open position).

7.4.2 Circuit Breakers

Section 7.3.6.3 discusses this electrical component.

7.4.3 Tripping Components

The term refers to all systems or components designed and used as safety devices and systems for opening and closing circuits in order to protect them from the various types of disruptions mentioned before. It is possible, for example, that tripping components are included in a circuit breaker.

7.4.4 Current Transformers

The current transformer (CT) is a type of transformer that reduces or multiplies alternating current. It is a type of "instrument transformer" that is designed to produce an alternating current in its secondary winding that is proportional to the current measured in its primary winding. A current transformer reduces high voltage currents to a much lower value and can be used to safely monitor the actual electrical current flowing through an AC transmission line using a standard current meter. The measuring instrument (low range AC ammeter) is connected to the terminals of the secondary winding. A current transformer is constructed in a similar manner to a conventional transformer. A transformer is fundamentally a step-down transformer (in terms of current), consisting of primary and secondary windings without any electrical connection. The number of turns in the secondary should be greater than the number of turns in the primary in order to step down current to a very low value.

7.4.5 Voltage Transformers

Voltage transformers are types of instrument transformers that reduce or step down the system voltage to a level that is measurable and safe. In comparison with the primary terminal, this transformer would produce a proportional amount of voltage at the second terminal. As a result, the primary windings of this transformer need to have more turns than those of the secondary windings in order to step down the voltage. The voltage transformer would convert the higher voltage into a lower voltage in a constant and linear manner.

7.4.6 Linear Couplers

The substation bus and switchgear are components of the power system that direct power to various feeders as well as isolate apparatuses and circuits from the power grid. These components include busbars, circuit breakers, fuses, disconnection devices, current transformers (CTs), voltage transformers (VTs), and the structures upon which they are mounted. In a power station, several alternators feed a common line called bus bar, or bus for short. In general, a busbar is a heavy conductor carrying thousands of amperes of electric current, as explained before. In order to isolate bus faults, all circuits connected to the bus are opened electrically by circuit breakers responding to relay action, direct-acting trip devices on low-voltage circuit breakers, or fuses. When the bus is disconnected, all loads and associated processes are shut off, and other components of the power system may be affected as well. It is extremely reliable to protect high-speed buses using the air-core current transformer or linear coupler method. The system can be easily expanded and modified in the future. In order to simplify switching circuits, the couplers can be open-circuited without any difficulty.

7.4.7 Transducers

As a sensor, the transducer detects abnormal conditions in the system and converts high values of short-circuit current and voltage to lower levels. There are two main types of sensors used: the current transformer and the potential transformer (PT). In a current transformer, there is a primary winding consisting of one turn and a secondary winding consisting of several turns. In order to maintain a constant secondary voltage, the potential transformer is designed to operate at a constant primary voltage.

7.5 ELECTRICAL GROUNDING SYSTEMS

Electrical systems and devices require proper grounding, or earthing, in order to function correctly. The terms ground, grounded, and grounding are frequently used in analyses of electrical installations. In various standards and codes, these terms are defined in a variety of formal ways. Nevertheless, grounding refers to the connection of the electrical system, electrical devices, and metal enclosures to the ground. In other words, it is the process of connecting to the earth, or earthing. The ground is a conductor of electricity, so we can refer to it as an electrical conductor. Furthermore, it is important to note that the earth is sometimes taken as a reference for voltage measurements since it has a potential of zero as a conductor. Due to the fact that most faults involve the ground, grounding of power systems is extremely important. Due to this, it plays an important role in both the protection of its components as well as the safety of its operators. Providing a low-resistance path for electricity to flow is the main purpose of grounding. Through the use of a grounding electrode, we can establish a connection to ground. In this manner, all the non-current-carrying conductors, such as the metal frame and housing of a computer, washer, dryer, electric drill, and so on can remain at a potential of 0 V. A variety of methods can be used to ground an electrical system to the ground, such as system grounding or neutral grounding. Neutral grounding refers to the connection of the neutral of the system or of the rotating system or transformer to the ground. Neutral grounding is an important aspect of power system design, since it has a significant impact on the performance of the system in terms of short circuits, stability, protection, and so on. In system grounding, the neutral conductor is intentionally connected to the earth. Lightning protection could be considered part of the grounding system. Lightning protection plays an important role in the design and operation of electric power systems. Lightning is the most common cause of outages and damage in areas with frequent storms. Through the use of down conductors to ground electrodes, lightning protection systems intercept or divert lightning and ensure that the surges are safely conducted to the ground by intercepting or diverting lightning. As a result, it helps prevent catastrophic events such as fires, injuries, and fatalities.

QUESTIONS AND ANSWERS

1. Identify the correct statement about power supply system.
   A. As a system, the power supply has the least impact on the system's reliability.
   B. Typically, remote oil and gas facilities generate power through the use of reciprocating engine generator packages or gas turbine generators.
   C. Generators that operate synchronously are also known as induction generators.
   D. Through the process of electromagnetic induction, a synchronous generator converts mechanical power into DC electric power.

   Answer) Option A is incorrect since the power supply system has the greatest effect on the reliability of the system. The correct answer is option B. As induction generators are asynchronous, option C is incorrect. Due to the fact that synchronous generators produce AC electricity, option D is incorrect.

2. It is a broad term that describes a wide variety of devices that all serve the same purpose: controlling, protecting, and isolating power systems.
   A. Transformer
   B. Busbar
   C. Switchgear
   D. Motor

   Answer) Option C is the correct answer.

3. Compared to cables with a similar cross-sectional area, these components can carry a greater amount of current.
   A. Busbars
   B. Switches
   C. Switchgears
   D. Transformers

   Answer) Option A is the correct answer.

4. What component is not considered part of an electrical protection system?
   A. Relay
   B. Circuit breaker
   C. Transformer
   D. Transducer

   Answer) Option C is the correct answer.

5. Regarding grounding systems, which statement is incorrect?
   A. Proper grounding, or earthing, is essential for the proper functioning of electrical systems and devices.
   B. When analyzing electrical installations, ground, grounded, and grounding are frequently used terms.
   C. Lightning protection is not considered grounding.
   D. Because earth has a potential of zero as a conductor, it may be used as a reference for voltage measurements.

   Answer) Option C is incorrect because lightning protection is considered a part of grounding.

FURTHER READING

1. Csanyi, E. (2016). *What is distribution substation and its main components?* [online] available at: https://electrical-engineering-portal.com/distribution-substation [access date: 1st August, 2023]
2. Csanyi, E. (2018). *The essentials of LV/MV/HV substation bus overcurrent and differential protection*. [online] available at: https://electrical-engineering-portal.com/substation-bus-overcurrent-differential-protection [access date: 8th August, 2023]
3. US Department of Energy. (2008). *Improving motor and drive system performance: A sourcebook for industry*. Washington, DC: US Department of Energy.

4. EEPOWER. (2020). *The basics of grounding electrical systems*. [online] available at: https://eepower.com/technical-articles/the-basics-of-grounding-electrical-systems/# [access date: 8th August, 2023]
5. Electrical Academia. (2023). *Power system protection components*. [online] available at: https://electricalacademia.com/electric-power/power-system-protection-components/ [access date: 1st August, 2023]
6. Worrell, E., & Galitsky, C. (2005). *Energy efficiency improvement and cost saving opportunities for petroleum refineries: An Energy Star® guide for energy and plant managers*. Berkeley: Energy Star, Ernest Orlando Lawrence Berkeley National Lab.

Index

L

M

N

O

P

W

Y

Printed in the United States
by Baker & Taylor Publisher Services